Disinfection Efficiency and Dose Measurement of Polychromatic UV Light

The mission of the Awwa Research Foundation (AwwaRF) is to advance the science of water to improve the quality of life. Funded primarily through annual subscription payments from over 1,000 utilities, consulting firms, and manufacturers in North America and abroad, AwwaRF sponsors research on all aspects of drinking water, including supply and resources, treatment, monitoring and analysis, distribution, management, and health effects.

From its headquarters in Denver, Colorado, the AwwaRF staff directs and supports the efforts of over 500 volunteers, who are the heart of the research program. These volunteers, serving on various boards and committees, use their expertise to select and monitor research studies to benefit the entire drinking water community.

Research findings are disseminated through a number of technology transfer activities, including research reports, conferences, videotape summaries, and periodicals.

Disinfection Efficiency and Dose Measurement of Polychromatic UV Light

Prepared by:
Karl G. Linden, Ph.D.
Duke University
Department of Civil and Environmental Engineering
121 Hudson Hall, Durham, NC 27708-0287

Alexander A. Mofidi, P.E.
Metropolitan Water District of Southern California
Water Quality Section
700 Moreno Avenue, La Verne, CA 91750-3399

Sponsored by:
Awwa Research Foundation
6666 West Quincy Avenue
Denver, CO 80235-3098

Published by the

Awwa Research Foundation
Advancing the Science of Water®

IWA Publishing

DISCLAIMER

Published by IWA Publishing, Alliance House, 12 Caxton Street, London SW1H 0QS, UK

Telephone: +44 (0) 20 7654 5500; Fax: +44 (0) 20 7654 5555; Email: publications@iwap.co.uk;
Web: **www.iwapublishing.com**

AwwaRF report number 90931F. Originally published by AwwaRF for its subscribers in 2003.
IWA Publishing version published 2004.

ISBN 1 84339 840 0

CONTENTS

TABLES

FIGURES

FOREWORD

The Awwa Research Foundation is a nonprofit corporation that is dedicated to the implementation of a research effort to help utilities respond to regulatory requirements and traditional high-priority concerns of the industry. The research agenda is developed through a process of consultation with subscribers and drinking water professionals. Under the umbrella of a Strategic Research Plan, the Research Advisory Council prioritizes the suggested projects based upon current and future needs, applicability, and past work; the recommendations are forwarded to the Board of Trustees for final selection. The foundation also sponsors research projects through the unsolicited proposal process; and various joint research efforts with organizations such as the U.S. Environmental Protection Agency, the U.S. Bureau of Reclamation, and the Association of California Water Agencies.

This publication is a result of one of these sponsored studies, and it is hoped that its findings will be applied in communities throughout the world. The following report serves not only as a means of communicating the results of the water industry's centralized research program but also as a tool to enlist the further support of the nonmember utilities and individuals.

Projects are managed closely from their inception to their final report by the foundation's staff and large cadre of volunteers who willingly contribute their time and expertise. The foundation serves a planning and management function and awards contracts to other institutions such as water utilities, universities, and engineering firms. The funding for this research effort comes primarily from the Subscription Program, through which water utilities subscribe to the research program and make an annual payment proportionate to the volume of water they deliver and consultants and manufacturers subscribe based on their annual billings. The program offers a cost-effective and fair method for funding research in the public interest.

A broad spectrum of water supply issues is addressed by the foundation's research agenda: resources, treatment and operations, distribution and storage, water quality and analysis, toxicology, economics, and management. The ultimate purpose of the coordinated effort is to assist water suppliers to provide the highest possible quality of water economically and reliably. The true benefits are realized when the results are implemented at the utility level. The foundation's trustees are pleased to offer this publication as a contribution toward that end.

There has been increasing interest in exploring the use of polychromatic, pulsed-ultraviolet (UV) and medium-pressure UV lamps as disinfectants for drinking water. However, there is currently no adequate basis upon which to evaluate the germicidal UV output from pulsed lamps so that appropriate comparisons can be made to medium-pressure and low-pressure UV lamps. To compare the three UV lamps, this report describes (1) the development of methodologies to measure the effective germicidal UV dose from polychromatic UV sources; and, (2) a comparison of the ability of the different types of UV lamps to inactivate heterotrophic plate count bacteria, MS-2 coliphage, and *Cryptosporidium parvum.*

Edmund G. Archuleta, P.E.
Chair, Board of Trustees
Awwa Research Foundation

James F. Manwaring, P.E.
Executive Director
Awwa Research Foundation

ACKNOWLEDGMENTS

The authors of this report are indebted to the following water utilities and individuals for their cooperation and participation in this project:

City of Burlington, Burlington, N.C., Stephen Shoaf
City of Wilmington, Wilmington, N.C., Mike E. Richardson
Metropolitan Water District of Southern California, Los Angeles, Calif., Dr. Mic Stewart

Research for this project headed by Professor Linden was carried out at the University of North Carolina at Charlotte and at Duke University. Work completed at both Universities could not have been possible without the help of students and laboratory staff. At UNC Charlotte Mr. Marvin Battle, a graduate research assistant and Ms. Shanshan Jin, a doctoral candidate, both in the Civil Engineering Department worked on the radiometry, UV dose modeling, and chemical actinometry portions as part of their graduate study. In addition, Scott Lane, an undergraduate research assistant, helped design, build, and test the bench scale UV units and assisted in literature development. Work completed at Duke University could not have been possible without the continued assistance of Ms. Shanshan Jin, then a doctoral candidate in the Department of Civil and Environmental Engineering at Duke University, who was responsible for performing most of the laboratory investigations relating to UV dose measurement using radiometry and actinometry. Dr. Charles Sharpless, a post-doctoral fellow, also provided assistance during the data analysis phase.

Work completed at the Metropolitan Water District of Southern California (Metropolitan) could not have been possible without the concerted efforts of many. Support from Metropolitan's Water Quality Laboratory is greatly appreciated. Support was provided by the Lab's Microbiology Development team members Hélène Baribeau, Ricardo De Leon, Anagha Chitre, Anne Johnson, Himansu Mehta, and Paul Rochelle (*Cryptosporidium* cell-culture infectivity, *Bacillus subtilis*, MS-2, and heterotrophic bacteria assays); Organic Chemistry Development team members Carrie Guo and Cordelia Hwang (uridine photohydrate analysis and quantitation); and, Process Development team members Brad Coffey (for document review and assistance in experimental plan development) and Connie Chou and Jerry Pitts (test sampling, chemical analyses, data collection, and research platform maintenance). Assistance from Bertrand Dussert (at the time, from Calgon Carbon Corporation) is greatly appreciated for arranging Metropolitan to use the collimated beam apparatus during this project.

Direction provided by the Project Advisory Committee—including Richard Haberman, California Department of Health Services, Fresno, CA; Jim Malley, University of New Hampshire, Durham, N.H.; and Joan Oppenheimer, Montgomery Watson, Pasadena, CA—and AwwaRF project officers Kim Hout Garrity and Albert Ilges is greatly appreciated.

EXECUTIVE SUMMARY

INTRODUCTION

Although ultraviolet (UV) irradiation for primary disinfection of surface or ground water supplies is not presently a widely accepted practice in the United States, there is currently intense interest in the use of this technology for helping to meet future regulations relating to improved microbial inactivation and decreased disinfection byproducts (DBPs). Much of this interest has been fueled by recent research indicating the effectiveness of UV technologies for the inactivation of *Cryptosporidium*. This scientific discovery, combined with proposed regulations concerning DBPs in the United States has resulted in a need to evaluate the feasibility and effectiveness of UV irradiation for use as a disinfectant for drinking water.

The interest in UV has had particular emphasis on the use of high-power, polychromatic, UV lamps. Although traditionally UV disinfection has employed lower intensity monochromatic (UV 254 nm) lamp sources (LP-UV lamps), the low intensity limits applications to smaller water treatment plants. The use of polychromatic, high-intensity, UV lamps may make practical the use of UV for large water treatment facilities. In addition to the lamp intensity issue, the interest in polychromatic lamps such as pulsed-UV (P-UV) and medium-pressure UV (MP-UV) lamps as disinfectants for drinking water was initially spurred by evidence of inactivation of *Cryptosporidium* using these light sources. Since these initial findings, it has also been shown that the traditional low pressure (LP) lamps are also very efficacious against *Cryptosporidium*. Currently, there is no adequate basis upon which to evaluate the germicidal UV output from the polychromatic, UV lamp sources so that appropriate comparisons of relative efficacy can be made between P-UV, MP-UV, and LP-UV lamps. To compare the three UV lamp technologies, this research was initiated to (1) develop and test methodologies to measure the effective germicidal UV dose from polychromatic, UV sources; and, (2) to compare of the ability of the different types of UV systems to inactivate microorganisms.

PROJECT APPROACH

The research was designed to evaluate the germicidal UV dose emitted from three different UV sources using dose measurement methods that included: radiometry, chemical actinometry, and biodosimetry. All work was performed on the bench scale utilizing systems designed to simplify the delivery of UV irradiation to a batch reactor such that accurate measurement of UV irradiance and dose could be made. The dose measurement methods were designed to evaluate the incident, average, and germicidal energy delivered to a water sample. Results from each measurement technique were compared and evaluated to determine the most appropriate means for measuring UV dose from polychromatic, UV lamps. Once the dose measurement methodology was established, the ability of LP-UV, MP-UV, and P-UV lamps to inactivate three different microbes was compared. The microorganisms tested included heterotrophic bacteria, MS2 coliphage, and *Cryptosporidium parvum* oocysts. The ability of heterotrophic bacteria to repair UV damage when exposed to light (simulated, uncovered reservoir storage) and dark (simulated, distribution system storage) environments was also evaluated.

KEY FINDINGS

Measuring UV dose on the bench scale was achievable with all the methodologies tested. Key to defining the delivered UV dose for polychromatic, UV lamps was defining a germicidal weighting factor for different wavelength emissions of UV light from the polychromatic sources. Because all wavelengths emitted in the UV range are not equally germicidal, a successful approach taken in this work was to define the DNA absorbance spectrum as the weighting factor to compare the germicidal dose from each light source.

Based on this germicidal dose weighting, it was determined that the inactivation of *C. parvum* was similar under all the UV lamp types tested. *Cryptosporidium* inactivation by UV light was rapid and reached 3 log inactivation at a dose of 10 mJ/cm^2.

The inactivation of heterotrophic bacteria was also similar for all lamp types and reached >2.5 log at a dose of 10 mJ/cm^2. After UV-treated bacteria were then incubated in environments which may simulate uncovered reservoir or distribution-system storage, significant increases in number of bacteria were observed. Bacterial recovery was attributed to both repair and regrowth. When 2.5 mg/L of combined chlorine was added after UV exposure, both repair and regrowth were suppressed. Thus a control strategy that may be implemented to prevent recovery of UV inactivated bacteria is chloramine addition.

It appeared that the inactivation of MS2 coliphage was appreciably improved when exposed to light from polychromatic, UV lamps as compared to the LP monochromatic UV lamp. A 4-log inactivation was achieved for P-, MP-, and LP-UV lamps at UV doses of 45, 60, and 70 mJ/cm^2 respectively. This difference was attributed to the fact that the germicidal weighting based on the DNA absorbance spectrum, may not be exactly representative of the nucleic acid absorbance spectrum for the MS2 coliphage, which has RNA as its nucleic acid. Thus the relative germicidal effectiveness of different wavelengths likely differs from the weighting of the DNA used for dose measurement.

CHAPTER 1
INTRODUCTION

BACKGROUND

Regulatory Impact on Drinking Water Treatment by UV Light

Ultraviolet irradiation for primary disinfection of surface or ground water supplies is not presently a widely accepted practice in the United States. However, recent research indicating the effectiveness of UV technologies for the inactivation of *Cryptosporidium* combined with proposed regulations concerning disinfection by-products (DBPs) in the United States has resulted in a need to evaluate the feasibility and effectiveness of UV radiation for use as a disinfectant for drinking water.

As UV technology develops and improves, the ability to benchmark the efficacy of new lamps must be maintained. Traditionally, an evaluation of UV disinfection is accomplished by determining the inactivation of an organism of interest as a function of UV dose applied. Therefore, to measure disinfection effectiveness, two things are necessary: (1) a means to measure the number of organisms before and after a treatment and (2) a means to measure the effective disinfection dose from that treatment. Microbiologists continue to develop methodologies for culturing many organisms of water quality importance and assess infectivity to enumerate these organisms. However, measurement of applied dose from the latest generation of UV technologies does not rely on any well-developed methodologies. New techniques for measuring the irradiance of interest (i.e., germicidal radiation) must be developed so that the next generation of UV disinfection systems can be adequately designed and the capabilities of the different types of UV technologies can be properly compared.

A recent development in the UV market that is of great interest to water utilities is the use of high-intensity, polychromatic, UV sources, such as medium-pressure UV (MP-UV) lamps and pulsed-UV (P-UV) lamps, for disinfection. These lamps are different from the familiar and traditional low-pressure UV (LP-UV) lamps. LP-UV lamps are easy to operate and monitor for germicidal irradiation output because they emit nearly monochromatic radiation at 254-nm, near the apparent peak wavelengths (260-265-nm) for germicidal effectiveness of many organisms. However, monitoring and measurement of germicidal irradiation from MP- and P-UV sources is challenging because the radiation emitted from these sources is spread throughout the germicidal range, up into the visible light range and beyond. An additional confounding factor is that the radiation emitted across the germicidal range is not equally effective at inactivating pathogens. One challenge is to develop, test, and utilize physical, chemical, and biological methodologies that can selectively measure the effective germicidal UV dose emitted from these sources and evaluate the disinfection efficiency of the new generation UV disinfection systems. Once the effective germicidal UV dose can be measured confidently, UV systems can be adequately designed and the relative effectiveness of each type of UV system can be rationally compared.

UV Technologies Overview

Although there are three principal UV technologies available on the market for disinfection, to date, the majority of the published literature on UV disinfection of drinking water

Table 1.1
Characteristics of typical low-pressure, LP-HO, medium-pressure, and pulsed-ultraviolet lamps

Characteristic	Low pressure / LP-HO	Medium pressure	Pulsed-UV
Emission	monochromatic (85-90% at 253.7 nm)	polychromatic (185-600 nm)	polychromatic (185 to 800)
Mercury vapor pressure (Pa)	0.1 to 10	50 to 300 kPa	N/A
Operating temperature (°C)	30 to 50 / 60 to 100	600 to 900	8,000-25,000 K
Arc length (cm)	15 to 200 / 15 to 200	10 to 200	10-100
Lifetime (hours in thousands)	8 to 12 / 7 to 10	3 to 8	>108 pulses
Efficiency (200-300 nm, %)	35 to 40 / 30 to 35	15-20%	<20% (varies)
Light irradiance (relative)	low / medium	high	high

has focused on the use of continuous wave UV irradiation from LP mercury vapor lamps (US EPA 1996a). Other UV technologies include continuous wave low pressure high output (LPHO) and MP mercury vapor lamps, and P-UV xenon arc lamps. The characteristics of typical LP-, LPHO-, MP-, and P-UV lamps are presented in Table 1.1. Comparison of the spectral outputs in the germicidal range of typical LP-, MP-, and P-UV sources are presented in Figure 1.1.

The environmental advantages of disinfection with LP-UV lamps, such as the lack of formation of harmful DBPs (Malley et al. 1995), are becoming more apparent, however, there are several disadvantages that make them impractical for some treatment facilities. High lamp numbers and large system footprints are sometimes required in facilities with effluent volumes higher than 10 million gallons per day (mgd) (Cairns 1996, Kwan et al. 1996). High maintenance and labor costs may also be associated with cleaning large numbers of UV lamps with sleeve fouling. Thus there may be interest in evaluating the use of higher irradiance UV sources such as LPHO-, MP- and P-UV lamps for larger scale drinking water applications.

Low-Pressure Ultraviolet Irradiation

LP- and LPHO-UV lamps characteristically emit monochromatic, low intensity radiation principally at 253.7 nm, within the germicidal wavelength range as defined by the DNA absorbance spectra. The germicidal UV dose (*H*) for these lamps is typically calculated as the product of the volume averaged incident irradiance (*E*, mW/cm^2) at 253.7 nm and the time of exposure (*t*, seconds) resulting in units of millijoules per square centimeter (mJ/cm^2) for UV dose.

Medium-Pressure Ultraviolet Irradiation

MP-UV lamps are mercury arc lamps that operate at up to a million times the pressure and at much higher temperatures compared to the LP-UV lamps. The increased pressure causes emission of higher intensity radiation and a broadening of emission lines, so that the emission becomes quasi-continuous and polychromatic over the range from the far UV (185-nm) to the infrared (1,367-nm). Approximately 15 to 25 percent of the polychromatic radiation emitted is in the germicidal range. Therefore, MP-UV lamps are less germicidally efficient than LP-UV

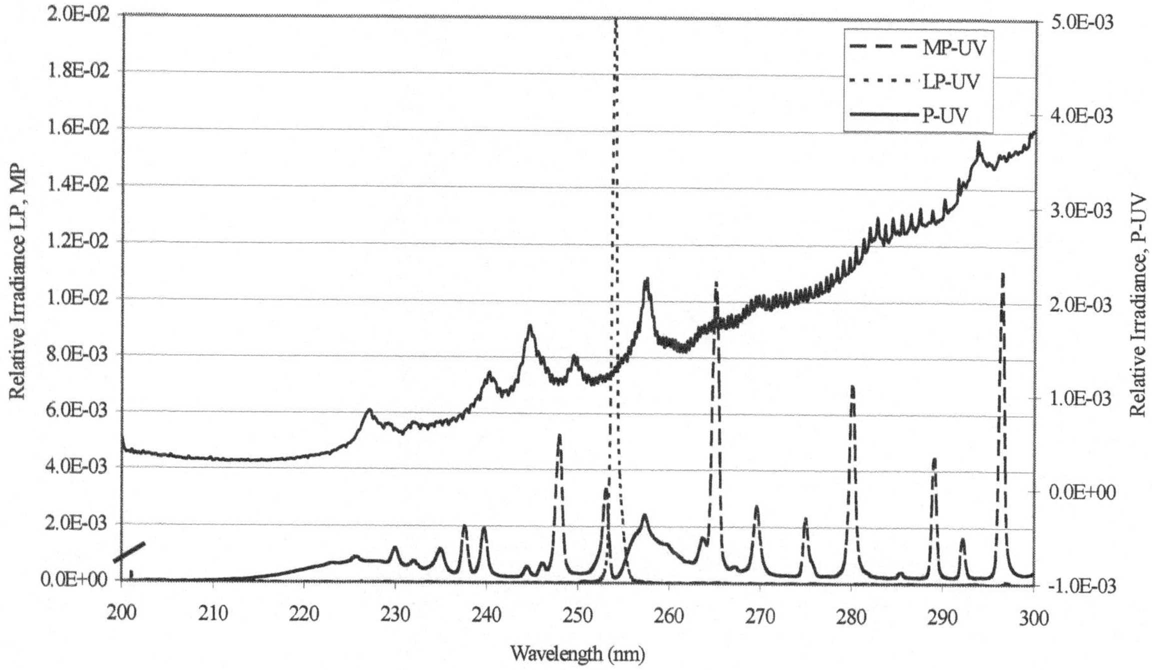

Figure 1.1 Relative irradiance in the germicidal ultraviolet range for low-pressure, medium-pressure, and pulsed-UV sources (wavelength output scaled relative to total 200-300 nm output). Note that low pressure high output lamps have an identical UV emission as the low pressure lamps.

lamps. However, because the radiation emittance is so much greater, MP-UV lamps can emit approximately 60 to 600 times the germicidal irradiance of the LP-UV lamps. Two of the strongest emissions in this range are at 266 and 256 nm. This higher emittance translates into a relatively smaller footprint for the MP-UV disinfection system (significantly fewer lamps for equivalent effect) and potentially lower operations and maintenance costs.

Pulsed-Ultraviolet Irradiation

The P-UV lamps (xenon, argon, krypton or other inert gases or mixtures) are characterized by a high irradiance broadband emission spectrum within 200-300 nm range. Their wavelength maximum and pulse rate can be adjusted electronically, making them of interest for various water treatment applications. P-UV treatment lamps are essentially "blackbody" emitters, providing continuous emission spectra between the wavelengths of 200- and 1,000-nm. High intensity, P-UV radiation is produced by accumulating electrical energy in an energy storage capacitor over fractions of a second and releasing this stored energy by a high speed electrical switch to do work in a few μs (Dunn et al. 1995). The result is extremely intense flashes of polychromatic radiation that are germicidally effective. Germicidal irradiance emitted by these lamps is not currently well defined. Recent interest has focused on the application to drinking water treatment. Prototypes of commercial systems are currently being researched for use in drinking water disinfection. However, because these systems are relatively new and most research has been proprietary, independent evaluations of system disinfection efficiency,

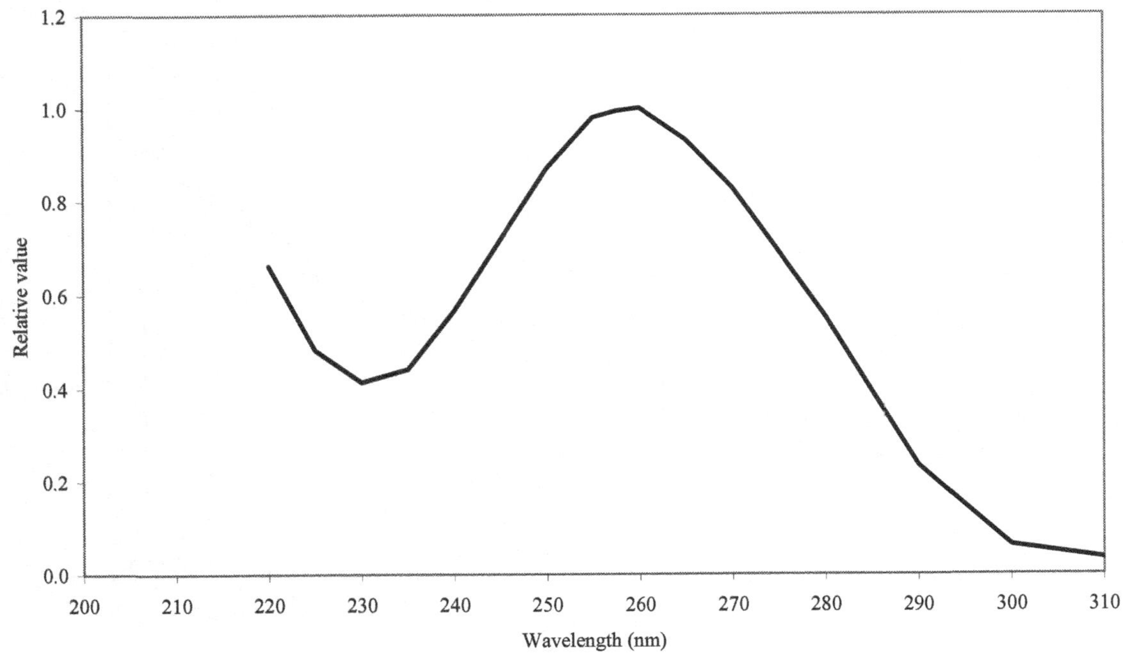

Source: Data from Jagger 1967.

Figure 1.2 Relative absorbance of DNA throughout the germicidal ultraviolet range (pH 7, 0.1M phosphate buffer)

mechanisms behind disinfection, and an evaluation of DBP formation potential must eventually be ascertained.

Measuring Disinfection Efficiency of Ultraviolet Lamps

When considering the calculation of UV dose as the product of irradiance and exposure time ($H = E \times t$), both irradiance and exposure time are complex entities that need further definition and evaluation. When considering measurement of germicidally effective UV irradiance, two principal factors must be examined: the UV source (e.g., mono/polychromatic, pulsed/continuous wave, LP/MP, lamp age and geometry/spacing/configuration), and the path to the target (e.g., water quality). The complexity of the calculation of UV irradiance in a disinfection system is dependent upon the relative complexity of these two factors in the system, described in detail below.

Ultraviolet Source

Historically, UV irradiance is a term used generically to describe the amount of radiation incident upon a target organism or compound. Because of the pervasiveness of the LP-UV lamp, the term UV irradiance has become synonymous with the idea of germicidal irradiance. This marriage of the two terms is not without merit when considering the monochromatic nature of the LP-UV source. The principal output of the LP-UV source is at 253.7 nm, near the peak of germicidal effectiveness as indicated by the absorption spectrum of DNA, illustrated in Figure 1.2. Therefore, measurement of UV irradiance, and subsequently germicidal irradiance, from LP sources is a relatively elementary matter.

The measurement of UV irradiance becomes considerably more complex when considering the MP-UV lamp which has a polychromatic output spanning the germicidal range and expanding into the visible and near infrared range. The amount of germicidal UV energy radiated from a MP-UV lamp may be considered as the total energy output by the lamp in the germicidal range. However, not all radiation line outputs in the germicidal range are equally effective at inactivating a target organism. For instance, in the case of DNA illustrated in Figure 1.2, the energy emitted at 260 nm is absorbed more efficiently than energy emitted at 290 nm. Because only energy that is absorbed can cause a photochemical or photobiological effect (the First Law of Photochemistry), the energy output at 260 nm will be more germicidally effective than that emitted at 290 nm. Therefore not all energy emitted in the germicidal range is equally effective for inactivation. The relative strength of different germicidal wavelengths may be considered analogous to the relative strength of OCL^- and HOCL chlorine species during chlorination. This point holds true for any microorganism: some wavelengths are more germicidally effective than others. The practical ramification is that when measuring the germicidal output of a polychromatic radiation source, each wavelength output must be weighted to its germicidal effectiveness to get a realistic representation of the amount of germicidally effective radiation that is being emitted. The value obtained using this method will be the actual germicidal UV irradiance. In addition, synergistic effects, enhancing inactivation, may occur from exposure to different wavelengths during irradiation with broad-band UV sources, further confounding the measurement of UV irradiance (Webb 1977).

A similar challenge exists when evaluating the polychromatic, P-UV source for germicidal irradiance output. An added complexity here is that the irradiance output from the pulsed source is not continuous, as is the case for the LP and MP mercury vapor lamps. Therefore, any measure of the irradiance must be linked to a time function describing the duration and frequency of pulses for an average detention time of the contact reactor.

Path to the Target

The path of UV radiation to the target consists of passage through the airspace within the quartz sleeve, the quartz sleeve itself, and the absorbing water column (and possibly blockage of light by particulate matter) to the target surface. The air within the quartz sleeve is considered to be non-UV absorbing although if ozone is produced some UV energy can be absorbed (US EPA 1986). The quartz sleeve itself absorbs less than 10 percent of the germicidal UV output. Of greater significance however, is the extent of fouling at the quartz/water interface. Fouling, often the result of chemical precipitation (scale) or biofilm growth on the quartz tube, can significantly interfere with the transmission of UV irradiance (Lin et al. 1997). Recent UV lamp designs incorporate an automated cleaning mechanism to minimize the buildup of scale and biofilm. For wastewater, the quality of the water through which the UV radiation must pass is often the most important factor in the path to the target. Water quality factors such as germicidal UV absorbance and turbidity, may attenuate UV radiation. In addition, particle size distribution may be an important monitoring tool to indicate the degree of particle association or particle shielding that is occurring in a water sample.

LITERATURE REVIEW

Disinfection Effectiveness of Ultraviolet Light

Bacteria

The efficacy of UV irradiation for inactivation of bacterial species has been investigated for many pathogenic and non-pathogenic strains. Based on a survey of data available in the literature the UV dose in mJ/cm^2 for a 2 log_{10} inactivation range from 1 to 12 and for a 4 log_{10} inactivation range from 3 to 34. Wilson et al. (1992) indicated that for a 4 log_{10} inactivation, *Aeromonas hydrophila, Campylobacter jejuni, E. coli* 0157:H7, *Legionella pneumoniae, Salmonella typhi, Shigella dysenteriae, Vibrio cholerae*, and *Yersinia enterocolitica* required respective doses of 5, 4.6, 5.6, 9.4, 8.2, 3, 2.9, and 4.6. Chang et al. (1985) found that *E. coli, Salmonella typhi, Shigella sonnei*, and *Staphylococcus faecalis* required UV doses of 8.4, 7.1, 8.2, and 11.2 for a 4-log_{10} inactivation. Studies of numerous *Salmonella* species by Tosa and Hirata (1998) (*S. anatum, S. derby, S. enteritidis, S. infantis, S. typhimurium*) indicated doses ranging from 5 to 15 for 3 or 4 log_{10} inactivation. Data for inactivation of heterotrophic plate count (HPC) bacteria indicate that these bacteria are no more resistant to UV than other bacteria. Typical UV doses required for a 3 log_{10} inactivation of indigenous (non-spiked) HPC are on the order of 10 mJ/cm^2 (with P-UV lamp) with <10 cfu/mL remaining after this dose (Mofidi et al. 2001).

Bacterial Reactivation

Because UV irradiation does not immediately result in cell lysis there is some opportunity for a UV inactivated cell to regain viability through a repair mechanism. One such repair mechanism is termed "photoreactivation" (Jagger 1958). Following UV inactivation, exposure to light in the 350 to 450 nm range (present in sunlight) could trigger an enzyme mediated process through which some bacteria may be able to repair the UV damage inflicted on the DNA and become viable. Not all bacteria possess this "photoreactivating enzyme" therefore not all bacteria can repair the UV damage. Furthermore, the ability for total coliform to photoreactivate is indirectly related to the applied UV dose. Lindenauer and Darby (1994) concluded that an increase in UV dose beyond that necessary for initial inactivation is valuable in minimizing photoreactivation events. As the applied UV dose was increased, the ability to recover decreased linearly up to a dose of approximately 140 where no reactivation was found. Similarly, a short exposure to photoreactivating light is less likely to result in significant repair taking place. Baron (1997) reported that a 1 log_{10} increase for the fecal streptococci should be designed for in wastewater systems operating at a dose of 40 mJ/cm^2. Similar magnitudes for photoreactivation were reported by Scheible et al. (1986). Most of the data available for photoreactivation have been collected for coliform organisms. However, organisms found to photorepair include total and fecal coliform, *E. coli, Streptococcus feacalis, Streptpmyces, Saccharomyces, Aerobacter, Micrococcus, Erwinia, Proteus, Penicillium,* and *Nuerospora* (Whitby et al. 1984, U.S. EPA 1986, Harris et al. 1987). Organisms demonstrating no tendency to photoreactivate include bacteriophage, somatic coliphage, *P. aeruginosa, Clostridium perfringens, Haemophilus influenzae, Diplococcus pneumoniae, B. subtilis*, and *Micrococcus radiodurans* (Harm 1980, Whitby et al. 1984, U.S. EPA 1986, Harris et al. 1987, Palmateer and Whitby 1987, Chrtek and Popp 1991).

Viruses

Viruses as a group appear to be the most UV resistant organism. Seven independent research groups (Chang et al. 1985, Harris et al. 1987, Wilson et al. 1992, Maier et al. 1995, Meng and Gerba 1996, Tree et al. 1997, Snicer et al. 2000) all reported a dose range of 20 to 30 mJ/cm^2 was required for a 4 $\log_{10}$ inactivation of Poliovirus Type 1. Results reported by Tree et al. (1997) were for wastewater with a turbidity of 20 NTU. Battigelli et al. (1993) reported that Coxsackievirus required a dose of 20 for a 3-$\log_{10}$ reduction, while Hepatitis A virus required a dose of 16 for 4-$\log_{10}$ reduction. Other researchers (Wilson et al. 1992, Wiedenmann et al. 1993, Snicer et al. 2000) reported similar results indicating that the Hepatitis A virus required a UV dose of between 7 and 30 mJ/cm^2 for 4 $\log_{10}$ reduction. Three research groups reported that Rotavirus strain SA 11 requires a dose of 23 to 26 mJ/cm^2 for 4-$\log_{10}$ inactivation (Wilson et al. 1992, Battigelli et al. 1993, Chang et al. 1985). Rotavirus strain WA was reported to require a dose of 50 mJ/cm^2 for 4 $\log_{10}$ reduction (Snicer et al. 2000). The most resistant virus type appears to be Adenoviruses. Meng and Gerba (1996) have reported that adenovirus exhibited significant resistance to UV, requiring doses of at least 120 for 4-$\log_{10}$ inactivation. These data were corroborated by other research groups (Thompson et al. 2000, Shin et al. 2001).

Protozoans

Previous studies based on viability assays such as excystation and dye inclusion/exclusion reported that UV doses greater than 8,000 mJ/cm^2 were necessary for a 2-3 $\log_{10}$ inactivation of *C. parvum* (Campbell et al. 1995). It was first determined in 1996 that P-UV radiation was effective for inactivation of *C. parvum* when assayed using a mouse infectivity endpoint (Arrowood et al. 1996). However, it was not until 1997 when Clancy and colleagues (Clancy et al. 1998a) reported that P-UV light was effective for *Cryptopsoridium* inactivation that this information was widespread in the water industry and since then numerous studies have followed to refine the UV dose - inactivation response for this pathogen. More than 6 independent data sets now exist for inactivation studies with *C. parvum* using either animal infectivity or cell culture infectivity endpoints. All the data conservatively suggest that a UV dose of 20 mJ/cm^2 will result in at least a 3 $\log_{10}$ inactivation of *C. parvum*. In fact, in many studies, a dose of 10 mJ/cm^2 resulted in the assay detection limit being reached.

P-UV work in 1996 (Arrowood et al. 1996) and 1997 (Clancy et al. 1998a, 1998b; Huffman et al. 2000) only investigated very high UV doses (>1,000 mJ/cm^2) and the dose measurement techniques were not validated substantially. This early work led to varying hypotheses that perhaps the pulsing nature and the extremely high energy of the light, or the polychromatic nature of the P-UV spectra was responsible for the susceptibility of *C. parvum*. These studies provided momentum for further investigation of *C. parvum* inactivation with other UV sources. Subsequent P-UV work by Clancy et al. (1998b) and Mofidi et al. (1998) illustrated that this technology also worked well at UV doses relevant to water treatment in tap and finished water respectively.

Bukhari et al. (1999) reported that MP-UV was effective at inactivating *C. parvum* up to 3.9 $\log_{10}$ inactivation at a UV dose of 19 mJ/cm^2 in a 200 gpm pilot system, using a mouse infectivity assay. Hargy (2000) reported > 4.5 log_{10} *C. parvum* inactivation in recycled filter backwash supernatant (NTU = 11) at a UV dose of 3 mJ/cm^2 on the bench scale. Hargy also tested recreational water at pilot scale with MP-UV at a dose of 10 mJ/cm^2 (40 gpm) and reported greater than 4.4 log_{10} inactivation. Subsequent MP-UV bench work by Craik et al. (2001) illustrated that a germicidal dose of <10 resulted in approximately 2.5 log_{10} inactivation. This MP-UV work, confirmed later by Mofidi and co-workers (2001), illustrated that a continuous wave UV source could perform as well as a pulsing source. Thus pulsing of UV was not necessarily the attribute that resulted in UV effectiveness.

UV irradiation from a monochromatic LP-UV source was investigated by Shin et al. (1999, 2001) and Clancy et al. (2000) for inactivation of *C. parvum* and illustrated that a dose as low as 4 mJ/cm^2 resulted in >3 log_{10} inactivation. Bench-scale research by Hargy (1999) also showed that a LP-UV dose of 10 mJ/cm^2 resulted in a 4 log_{10} inactivation. Gillogly presented similar results of inactivation of *C. parvum* in waters containing 2 and 7 NTU and found no difference in inactivation relative to the level of turbidity (Stolarik et al. 2001). Thus monochromatic UV appears to be as efficient at inactivation of *C. parvum* as MP- and P-UV sources.

In a controlled study of the potential for *C. parvum* to repair via dark or light mediated pathways, there was no evidence of any reactivation in multiple trials (Shin et al. 2001). Reactivation conditions tested were optimized for both prokaryotic and mammalian repair and recovery. UV doses used for inactivation prior to reactivation attempts were low, approximately 1 to 2 mJ/cm^2, and resulted in approximately 0.75 to 2 log_{10} inactivation. The low doses and optimum reactivation conditions were chosen to allow the best chance for reactivation to occur. However, no increase in survivors following reactivation protocols was evident. An Awwa Research Foundation project currently underway (Rochelle, P.A., M.M. Marshall, A. Mofidi, B. Montelone, and S.J. Upton. 2000. AwwaRF project # 2669: An Investigation of Cryptosporidium parvum Oocyst Repair Following UV Disinfection) has found 11 repair genes present in *C. parvum*. Although the genes seem to be dormant in *C. parvum* (no oocyst repair has been seen after UV exposure), experiments have involved sequencing, cloning, and placing these genes in yeast cells to determine if they can become active.

Similar to *Cryptosporidium*, initial studies of the UV inactivation of *Giardia* was performed using viability assays (excystation) and reported UV doses were relatively high and considered impractical for use of UV for *Giardia* control. Karanis et al. (1992) Carlson et al. (1985) and Rice and Hoff (1981) all reported UV doses necessary for 1 to 2 log_{10} inactivation at 60 to 120 mJ/cm^2 based on excystation viability assays. Although these doses are not extremely high, they were considered impractical when *Giardia* could be effectively controlled using moderate doses of chlorine.

UV disinfection of *Giardia lamblia* (human pathogen) has been investigated by at least four independent research groups utilizing an animal model (mongolian gerbils) for infectivity assays. Linden et al. (2002) and Mofidi et al. (2002) reported that in repeated trials, a dose of approximately 4 mJ/cm^2 of LP-UV resulted in >3 log_{10} reduction in infectivity of *Giardia lamblia* in lab water and filtered water, respectively. Furthermore, at UV doses of 16 and 40 mJ/cm^2 Linden et al. (2002) reported that neither light nor dark repair protocols resulted in any reactivation of *G. lamblia* as evidenced by no changes in infectivity. Malley (2000) reported that light from a P-UV source achieved greater than 5 log_{10} inactivation of *Giardia lamblia* in secondary effluent and biological nutrient removal treated wastewater at a UV dose of approximately 8 mJ/cm^2.

Ultraviolet Dose Measurement

A number of techniques have been developed to measure UV irradiance or UV dose directly for use with bench and full scale reactor systems. Current accepted techniques for measuring UV irradiance/dose include physical, mathematical, chemical, and biological approaches. These techniques are documented in the literature (Qualls and Johnson 1983, US EPA 1986, US EPA 1996a, Linden and Darby 1997). A brief review of dose measurement at the reactor scale is presented followed by bench scale dose measurement techniques.

Reactor-Scale UV Dose Measurement

Physical techniques. Physical methods principally involve radiometry, using UV sensors calibrated to narrow or broad-band spectral responsivities. These sensors can be used to crudely estimate the UV dose in a reactor but only provide information on UV irradiance at one point in a reactor and provide no information on hydraulic retention time. Sensors alone are not considered appropriate for reactor scale dose measurement. Because most sensors can only accurately measure radiation irradiance incident normal to the plane of the detector, these instruments are principally used for relative UV measurements. Sensor based radiometry has been employed routinely in full scale reactors for monitoring constancy of UV output and to indicate when irradiance has decreased due to fouling or aging of the lamps.

Mathematical techniques. The Point Source Summation (PSS) method is a mathematical technique commonly employed for estimating UV irradiance in LP-UV systems, although other methods have been proposed (Darby et al. 1993). The principal drawback of the PSS technique is that it only provides a single averaged dose value, whereas the actual dose that a target receives will vary according to its flow path through the reactor. Furthermore, the determination of detention time in most UV systems assumes that plug flow conditions exist, an assumption that must be tested on a case by case basis. Blatchley and co-workers (Chiu et al. 1999) have developed computational fluid dynamics (CFD) models that utilize a random walk approach to track particles and develop a dose distribution within LP-UV reactors (Chiu et al. 1999). When employing mathematical models, measurement of UV absorbance must account for particle scattering and absorption (Linden and Darby 1998) and the accuracy of lamp spectral output for polychromatic sources must be confirmed, preferably independent of the manufacturer, including prediction of changes in lamp output due to lamp aging. Changes in the absorbance of the UV quartz sleeve due to fouling must be accurately monitored or modeled (Lyn et al. 1999). Mathematical models for use with polychromatic UV sources are currently under development (Bolton 2000a).

Chemical techniques. Chemical actinometry is the measurement of a radiation dose by the change produced in a chemical. Similar to the mathematical approach, chemical actinometry when applied to the bulk liquid flow can provide a single UV dose value to the whole complex system. Although use of this type of actinometry application may be appropriate for one time testing and evaluation of UV systems, it would not be feasible for routine monitoring of UV dose. Application of actinometry to measure dose in operating full-scale UV disinfection systems necessitates innovation in how the actinometer is introduced into the reaction chamber. Development of some type of carrier system that separates the actinometer from the bulk liquid flow, yet allows estimation of a realistic dose or UV irradiance within the reactor, is necessary.

Biological techniques. Biological measurement of dose has centered on the bioassay, (also termed biodosimetry) technique. Because it is a biologically based test, the method provides a measure of the effective germicidal dose - the dose necessary for a desired germicidal effect. Similar to the mathematical and actinometric approaches, this method provides a single average dose value to the whole complex system. This technique does not provide any information as to the dose experienced by a clumped or floc-enmeshed organism, nor does it fully characterize the dose provided at different flow rates other than those tested. Sommer and Cabaj (1993) identified four criteria for selection of a bioassay test organism including: (1) a steady inactivation over a wide dose range, (2) a simple and sufficient culturability, (3) non-pathogenic, (4) a temporal stability with regard to UV susceptibility. Typical organisms that have been used with bioassays include: *B. subtilis* spores (Sommer and Cabaj 1993, Qualls and Johnson 1983), and MS2 coliphage (Braunstein et al. 1996). Because this is a biologically based approach, the inactivation of the target organism must be assumed to be solely caused by the effects of the magnitude of UV dose exposure and that no alternative reactions, chemical or otherwise, are contributing to the inactivation. Application of the bioassay technique to UV systems where synergistic reactions could be occurring because of photosensitized compounds may result in miscalculation of actual radiation dose values.

Bench-Scale UV Dose Measurement

When performing bench scale dose-response studies, care must be taken to accurately characterize the UV irradiance. This must include a thorough description of how irradiance was determined, especially for polychromatic UV sources.

Radiometry methods. Radiometers are typically used to measure irradiance on the bench scale. Some important steps to take in the proper characterization of irradiance are presented below.

Petri factor determination. Because radiation from a collimated beam is not actually 100 percent collimated, the irradiance measured at the center of the surface area of a petri dish is corrected to represent the average irradiance across the petri dish surface area. This type of correction must be made when performing bench scale work. Dose calculations made using irradiance measurements solely from the center of the Petri dish may seriously confound subsequent results.

Radiometer characterization. There are many types of radiometers available for use. However, it is essential that the responsivity characteristics of a radiometer be well defined when using a radiometer to estimate irradiance from a polychromatic source. The irradiance may be representative of the total energy in a predefined "germicidal range" such as 200 to 300 nm or may represent a germicidal range irradiance weighted to the relative effectiveness of each specific wavelength.

One radiometer available for UV measurement (International Light, IL1770, SED240 detector, W diffuser, germicidal [G] filter) utilizes a germicidal filter that internally weighs the detection of a polychromatic spectra to provide direct measurement of effective germicidal radiation, spectrally weighted to match the IES (Illumination Engineering Society) Luckiesh and German DIN (Deutsche Industrialen Normalen) standard germicidal effective irradiance curves. The response of the radiometer with the G filter as compared to these germicidal effective irradiance standards is illustrated in Figure 1.3. This radiometer can, in theory, be used to directly read the effective germicidal radiation from a polychromatic source such as the MP-UV and P-UV

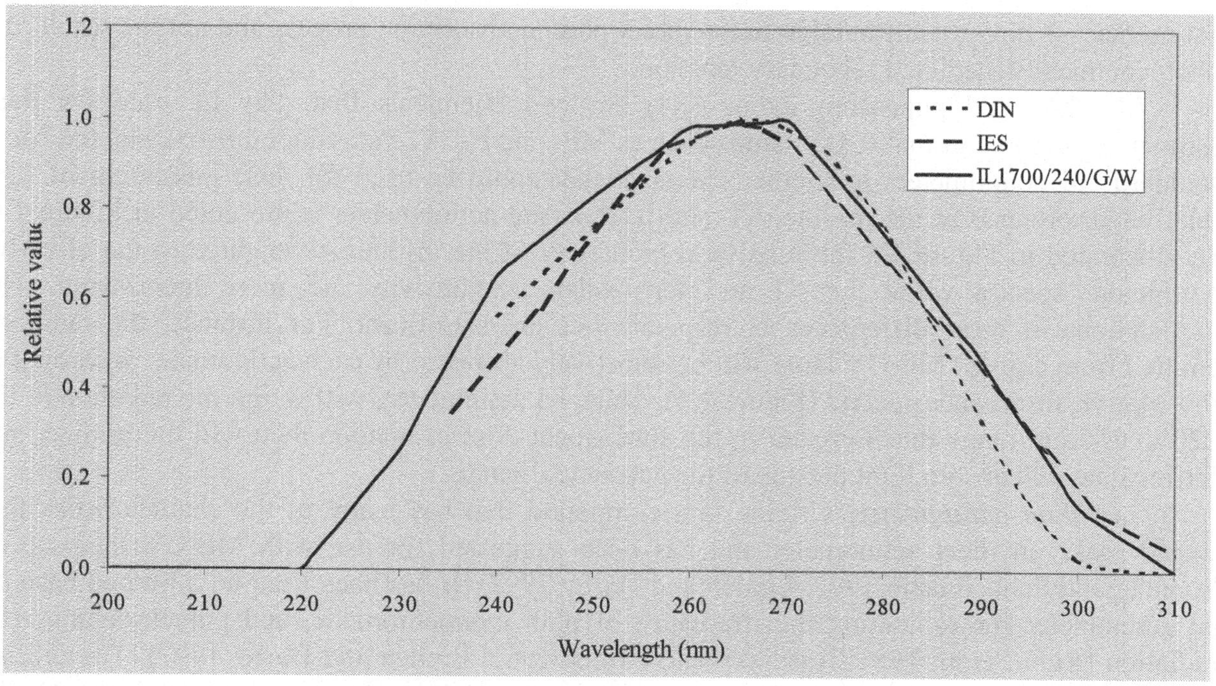

Figure 1.3 Comparison of the German DIN and the IES standard germicidal effective irradiance curves with the relative responsivity of the IL 1700, SED240/G/W germicidal radiometer

sources. When measuring UV irradiance with a monochromatic LP-UV source, the sensitivity factor for UV-254 nm is enabled to provide calibrated irradiance readings. An estimation of the actual UV irradiation in the 220 to 300 nm range can be accomplished from the detailed calibration data along with the emission spectra of the MP-UV lamp (Linden and Darby 1997).

The same detector can be used without the G filter to read a broadband UVC output, unbiased by the G filter but with its own internal responsivity that varies across the UVC range. However, to use this type of detector, knowledge of the relative spectral output of the UV lamp in use is necessary, as well as characterization of the responsivity of the UV detector. Ideally, radiometers utilized for polychromatic UV measurements should be detailed for responsivity factors across the wavelength range of interest to make the proper corrections for irradiance estimations.

Chemical actinometry. Many chemical actinometers can be used to measure radiation at wavelengths that span the absorption spectrum of nucleic acids in the far UV regions (Blaney et al. 1993, Gahr et al. 1995). However, most of the well-researched actinometers also absorb radiation at wavelengths outside the germicidal range and the reported quantum yields (moles of chemical changed per Einstein of photons absorbed) are not constant over a wide wavelength range. Ideal actinometers for measuring polychromatic, germicidal radiation would have the following characteristics (von Sonntag and Schuchmann 1992): (1) absorb only germicidal radiation, (2) absorbance spectra similar to that of DNA, (3) constant quantum yield throughout germicidal wavelength range, (4) easy to quantify using standard laboratory equipment, (5) non toxic in the environment. With any chemical actinometer, care must be taken when evaluating it for use with water sources of a diverse chemical nature to assure that non-photochemically induced degradation of the actinometer is not a factor. Similarly, actinometry with polychromatic and/or high irradiance sources must validate that the change produced in the chemical actinometer is directly attributable to the direct photon absorption process and not the result of a photochemically mediated secondary reaction.

Uridine and potassium iodide (KI) are two chemicals that may be used for dose estimation of polychromatic, UV lamps such as MP- and P-UV. Potassium ferrioxalate, the "gold standard" of actinometers is another chemical that could be used for dose measurement. The relative absorbance of the uridine, KI, and ferrioxalate actinometers is presented in Figure 1.4. As illustrated in Figure 1.4 the relative responsivity of the uridine is weighted to the effective germicidal spectra while the KI and ferrioxalate responsivity are more broad-band. The implications of these differences in responsivities are significant. For instance, the radiation emitted from a given MP-UV lamp will be selectively absorbed by each actinometer according to the relative absorbance spectra (Figure 1.4). Thus, KI actinometry will weigh the radiation in the 220 to 250 nm range much greater in the subsequent dose estimation than will the uridine, and ferrioxalate will absorb light outside of the germicidal range.

Uridine actinometry. Uridine is a compound that has many of the characteristics that would make an ideal actinometer and has been suggested for use with MP-UV lamps (von Sonntag and Schuchmann 1992, Linden and Darby 1997). It has been used on a limited basis as an actinometer for estimating the irradiance of both monochromatic, and polychromatic, UV radiation (Rubin et al. 1981, Rushizky and Pardee 1962, Linden and Darby 1997). The uridine molecule contains uracil, a naturally occurring nucleobase, as its chromophore (See Figure 3.14). The absorbance of uridine, expressed as the absorbance relative to the peak absorption at 262 nm, is illustrated in Figure 1.4. Upon UV irradiation, uridine undergoes a 5', 6' photohydration reaction, whereby a water molecule replaces the 5', 6' double bond, forming a photohydrate product. This photohydrate product does not absorb radiation at 262 nm (Moore and Thompson, 1955). The quantum yield for the photohydration reaction of uridine is relatively constant in the wavelength range of 238 to 280 nm (Swenson and Setlow 1963). In addition, the photohydration reaction was reported to be independent of wavelength between 240 to 280 nm (Rahn and Stelin 1979). Because only absorbed radiation can produce a chemical change, the degradation of uridine is expected to have a wavelength dependence similar to its absorbance spectrum (Swensen and Setlow 1963). The presence of uridine can be monitored through spectrophotometric analysis of the compound absorption peak at 262 nm. UV dose can be estimated using uridine from an analysis of the change in absorbance at 262 nm before and after UV exposure.

Potassium iodide actinometer. The potassium iodide (KI) actinometer is based on the photolysis of a mixture of potassium iodide and iodate (Rahn 1997). This actinometer can be modified for use with any UV lamp. Iodide exposed to ultraviolet light while in aqueous solution undergoes a photoionization reaction with a low quantum yield due to most aqueous electrons recombining with the iodide atoms in the solvent. Also, iodide (even as a 0.6 M solution) absorbs only out to 275 nm. Therefore, a KI-only based solution would be a poor UV actinometer. When iodate (IO_3) was added to a KI solution, the IO_3 acted as an electron acceptor and formed a complex with the KI having two effects: (1) the absorption spectrum is shifted to the red and (2) the quantum yield increases to 0.75. The reaction generates triiodide ion (I_3^-) which absorbs only at 352 nm. Spectrophotometric analysis of solution absorbance at 352 before and after exposure to UV radiation can quantify the photon flux, and thus, the UV dose. Stability problems with this actinometer are discussed in Chapter 3.

Potassium ferrioxalate actinometer. The potassium ferrioxalate actinometer is based on the photolysis of ferric iron [III] and subsequent reduction to the ferrous [II] state. Ferrous iron is

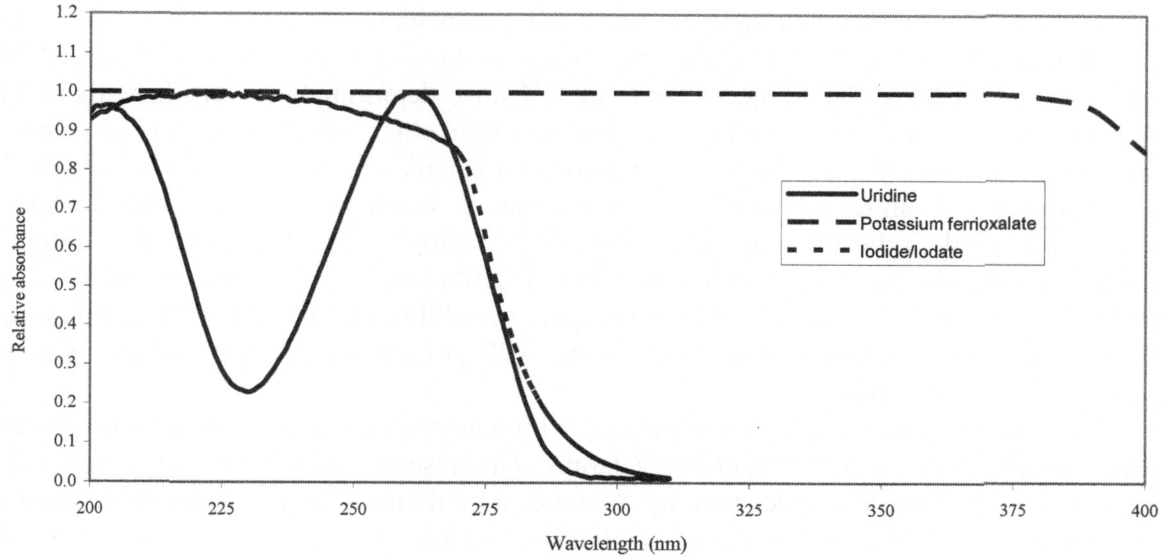

Figure 1.4 Comparison of the uridine, iodide/iodate, and ferrioxalate actinometer absorbance curves

measured before and after UV exposure and the absorbed UV dose is quantified using the reaction quantum yield (Jagger 1967). The molar concentration of ferrous iron is monitored via a chemical reaction with phenanthroline and absorbance measurement at 510 nm.

Biological Actinometry

Biological actinometry (also referred to as biodosimetry or bioassay), although primarily used for estimating the UV dose in a flow through reactor system, can also be used to measure UV irradiance or dose from polychromatic, UV lamps where the measurement of dose is not well defined. UV dose measurement can be performed by employing a well studies microorganism such as *B. subtilis* spores. The spores may be exposed to "collimated" radiation from a LP-UV lamp from which dose is accurately determined using a radiometer. A relationship between applied UV dose and inactivation of the test organism can be developed with the LP-UV lamp. Spores may then be irradiated by the polychromatic, UV set-up and enumerated before and after UV exposure. The magnitude of inactivation provided by the polychromatic lamps can be used to back-calculate the applied dose from the polychromatic lamps using the applied dose/log reduction relationship developed for the easy-to-measure monochromatic lamp. This relatively simplified approach to dose estimation for "unknown" UV lamps such as MP- and P-UV lamps may provide a comparative basis for investigating microbial inactivation among different UV lamp types.

STUDY SCOPE AND RESEARCH OBJECTIVES

The goal of this research was to evaluate the disinfection effectiveness of polychromatic, UV radiation from MP and P-UV lamps. To accomplish this, the project was organized to develop and evaluate methodologies that could be used to measure the effective germicidal UV dose from a polychromatic, UV source. Once the germicidally effective UV dose could be accurately estimated from the polychromatic source, a fair comparison to LP-UV disinfection effectiveness could be made. Three methods of UV dose determination were investigated and applied to LP, MP, and P-UV lamps. Specific research objectives were to (1) develop and evaluate physical, chemical, and biological methods for calculating effective germicidal UV dose from MP and P-UV lamps, (2) establish a UV dose/log inactivation relationship for specific bacterial and viral indicators for MP and P-UV lamps, (3) determine the extent of photoreactivation and dark repair (recovery from inactivation) of heterotrophic bacteria after treatment by MP and P-UV lamps, and (4) compare the ability of MP and P-UV lamps in their ability to inactivate *Cryptosporidium*. All testing with polychromatic light sources was done side-by-side with a LP lamp.

The dose estimation work was combined with inactivation experiments to benchmark the disinfection efficiency of the MP and P-UV lamps. This research was organized to be a unique step toward a more complete understanding of and design of emerging UV technologies that will likely be used by drinking water treatment plants in the near future. To accomplish this, dose measurement expertise at Duke University was combined with process development expertise at Metropolitan. Specific tasks of this project are summarized below.

Development of Methods to Calculate Germicidal UV Dose

This phase of research employed physically simplified systems that used LP, MP, and P-UV lamps in a "collimated beam" mode to produce quasi-parallel light. UV dose was measured using three techniques: radiometry, bioassay, and chemical actinometry described below.

Radiometry

A radiometer (IL 1700; International Light Inc., Newburyport, Mass.) equipped with a detector and filter (SEL240/W) was used to measure the irradiance from the LP monochromatic source at 253.7-nm. The IL 1770 germicidal radiometer system equipped with a detector and germicidal filter (SEL240/G/W) providing direct measure of the effective germicidal radiation spectrally weighted to match the IES Luckiesh and DIN standard germicidal effective irradiance curves, and was used to estimate the germicidal irradiance from both the MP- and P-UV light sources. Radiometer and detector were also utilized without the germicidal filter.

Radiometric Mathematical Model

In addition to the germicidal detector, a mathematical model was refined and evaluated to determine the effective germicidal dose from the polychromatic lamp sources. The model was based on irradiance values from a calibrated radiometer and detector, the absorbance characteristics of the water/microorganism suspension, the emission output at each wavelength for a polychromatic, UV lamp such as the MP and P-UV lamps, and the relative sensitivity of the detector used to measure the UV irradiance. The dose results from the mathematical model were correlated with the other dose estimation methods (Luckiesh/DIN germicidal detector, chemical actinometry, and biodosimetry).

Biological Actinometry

Biological actinometry for polychromatic dose measurement employed *B. subtilis* spores. Spores were exposed to "collimated" radiation from a LP-UV lamp from which dose was accurately determined using a radiometer. A relationship between applied UV dose and inactivation of the test organism was developed with the LP-UV lamp. Spores were then irradiated by the polychromatic, UV lamps and enumerated before and after UV exposure. The magnitude of inactivation provided by the polychromatic lamps was used to back-calculate the applied dose provided by the polychromatic lamps using the applied-dose, log-reduction relationship developed for the easy-to-measure monochromatic system. Although it is a relatively simplified approach to dose estimation, biodosimetry provides an accepted, comparative basis for investigating microbial inactivation from UV lamps.

Chemical Actinometry

Two chemical actinometers were evaluated as part of this research effort. Uridine, a derivative of the uracil nucleobase, was used as a chemical actinometer to estimate UV dose. The degradation of uridine was monitored through spectrophotometric analysis of the uridine absorption peak at 262-nm and development of a calibration curve in the absorption range of interest. Because the quantum yield of uridine is based on a photohydration reaction being the sole photodegradation product, an assurance was needed to confirm this when used for the polychromatic lamps. It was postulated that the high intensity of the P-UV lamp may lead to formation of other degradation products. Therefore it was necessary to analytically determine the types of degradation products or by-products that are formed.

In addition to uridine, it was decided to add the potassium iodide/iodate actinometer as an additional chemical used to estimate UV dose. The decay of the iodide/iodate actinometer was monitored through spectrophotometric analysis of the photoproduct at 352 nm. However, no detailed analysis of the photoproducts was performed due to budget and time constraints.

Establish UV Dose/Log-Inactivation Relationships

This phase of research established the efficiency of each UV lamp in its ability to inactivate heterotrophic bacteria, and MS-2 bacteriophage. The organisms were suspended in a petri dish (without cover) and exposed to a series of increasing doses at a minimum of 5 dose steps. The MS-2 was prepared in buffered distilled water and heterotrophic bacteria was from biofilter effluent water. The samples were approximately 1 cm deep and continuously mixed using a micro stirbar and magnetic stirrer. Lamp UV irradiance was measured by use of the radiometry and actinometry methods developed during the study. These tests were conducted in a filtered natural water matrix.

Evaluate Photoreactivation and Dark Repair of Heterotrophic Bacteria

Photoreactivation studies were designed to provide an assessment of the photoreactivation (light mediated) and dark repair (repair in the absence of light) capabilities of heterotrophic bacteria following UV treatment. Testing involved exposing samples of UV-treated water to both light and dark conditions over select time periods at constant temperature, representative of stored, treated water.

Quantify *Cryptosporidium* Inactivation by Ultraviolet Light

After developing a fundamental understanding of dose measurement and UV-dose, log-inactivation and inactivation/reactivation relationships, a focused evaluation of the inactivation of *C. parvum* was conducted. This included LP, MP, and P-UV disinfection of *C. parvum* with an emphasis placed on comparing results between the lamp technologies. *C. parvum* oocysts were assayed by a human cell-culture infectivity assay described by Rochelle et al. (1997) and Mofidi et al. (2001).

KEY PROJECT QUESTIONS

From these objectives, several key questions were developed to help guide this research effort:

- What dose measurement techniques are most appropriate for use with polychromatic lamps?
- Is UV irradiation from different UV lamp sources equally germicidal?
- To what extent can heterotrophic bacteria repair UV damage or regrow after UV exposure?
- Is *Cryptosporidium* equally susceptible to UV irradiation from different UV lamps?

PROJECT TEAM

This project was a joint effort between Duke University and Metropolitan Water District of Southern California. In total, three utilities made in-kind contributions to the project by supplying water for bench studies investigating the efficiency of UV in different sourcewaters. The contributing utilities were:

- City of Wilmington, NC
- City of Burlington, NC
- Metropolitan Water District of Southern California, La Verne, CA

CHAPTER 2
MATERIALS AND METHODS

EXPERIMENTAL SETUP

Low Pressure Ultraviolet Lamp Units

Bench-Scale Setup

The LP-UV lamp (GPH463T5L/4/UVC/2, 30 W; Atlantic Ultraviolet Corp., Hauppauge, N.Y.) was mounted in a bench scale unit typically referred to as a “collimated beam” set-up, provided by Calgon Carbon Corporation Inc., illustrated in Figure 2.1. The unit consisted of a lamp, lamp housing, power supply, 6.4-cm (2.5-in.) inside diameter “collimating” tube, and pneumatically controlled shutter.

Lamp Output Spectrum

The spectrum for the LP-UV lamp, presented in Figure 2.2, was evaluated with an Ocean Optics S2000 (Dunedin, FL) spectrometer. Of interest here is the sheer prominence of the peak emission at 254 nm. There is some output at other wavelengths, including two of the peak emissions from household fluorescent lamps (436 and 546 nm).

When monitoring the irradiance of LP-UV lamps with radiometry, as long as the detector is only sensitive to the germicidal range, the total irradiance recorded will be that at 254 nm. However, when utilizing broadband actinometers, such as the potassium ferrioxalate actinometer, to get a value for UV irradiance, the irradiance from wavelengths outside of the germicidal range needs to be subtracted. This problem would not be encountered with the uridine and iodide/iodate actinometers because they do not appreciably absorb light above 300 nm (See Figure 1.4).

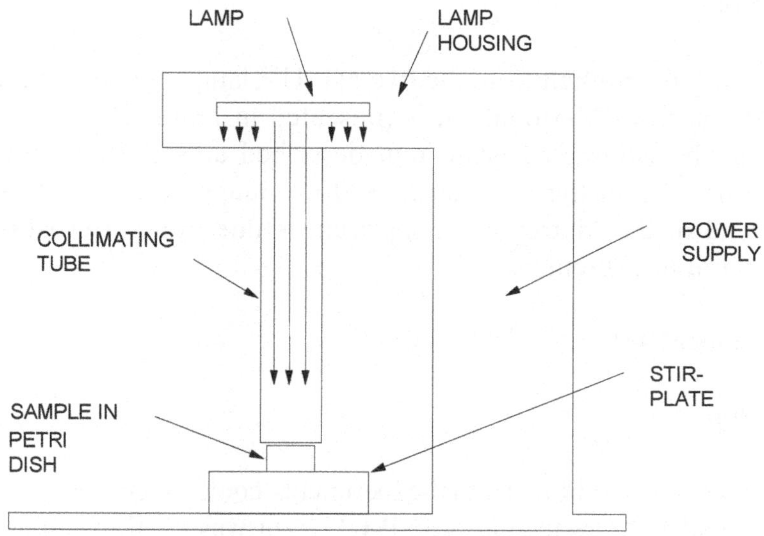

Figure 2.1 Illustration showing low- and medium-pressure “collimated beam” unit

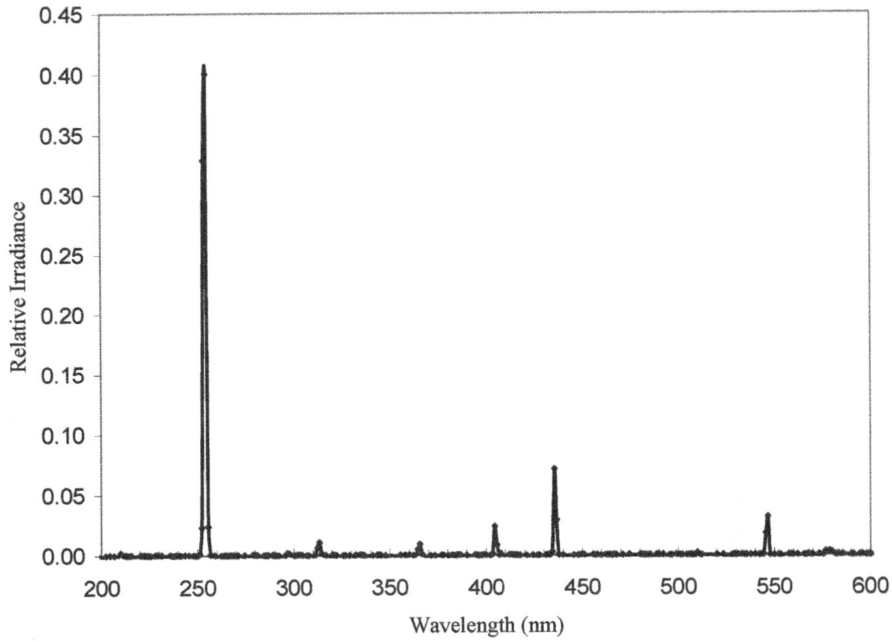

Figure 2.2 Emission spectra from a 15-W, low-pressure, ultraviolet lamp

Medium Pressure Ultraviolet Lamp Units

Collimated Beam Setup

The bench scale MP-UV system used during this research was identical to the system described for the LP-UV system (Figure 2.1) but utilized a MP-UV lamp instead of the LP-UV lamp. Two different collimating tube lengths were evaluated for the MP system, 20 in. (51 cm) and 50 in. (127 cm), to allow for higher and lower irradiance positions, respectively.

Lamp Output Spectrum

The spectrum of light emitted from the two MP-UV lamps utilized in this study, one from Duke University and one from Metropolitan is presented in Figure 2.3. Both lamps are Rayox 1kW lamps housed in the collimated beam unit described above. Both lamps exhibit similar wavelength emission profiles in the UV 200 to 300 nm range. The "noise" in the wavelength profile at the lower end of the Metropolitan spectrum is due to a perturbation in spectrometer setting during the spectral measurement.

Pulsed-Ultraviolet Lamp Unit

Collimated Beam Setup

The P-UV system was modified so that experiments could be conducted in a manner similar to that for the LP-UV and MP-UV testing. The P-UV light was "collimated" with a 3-in. (8-cm) diameter PVC tube to expose 60-mm diameter petri dish samples at least 15-in (38-cm) away. The original side-port exposure and collimated-exposure configurations are illustrated in Figure 2.4.

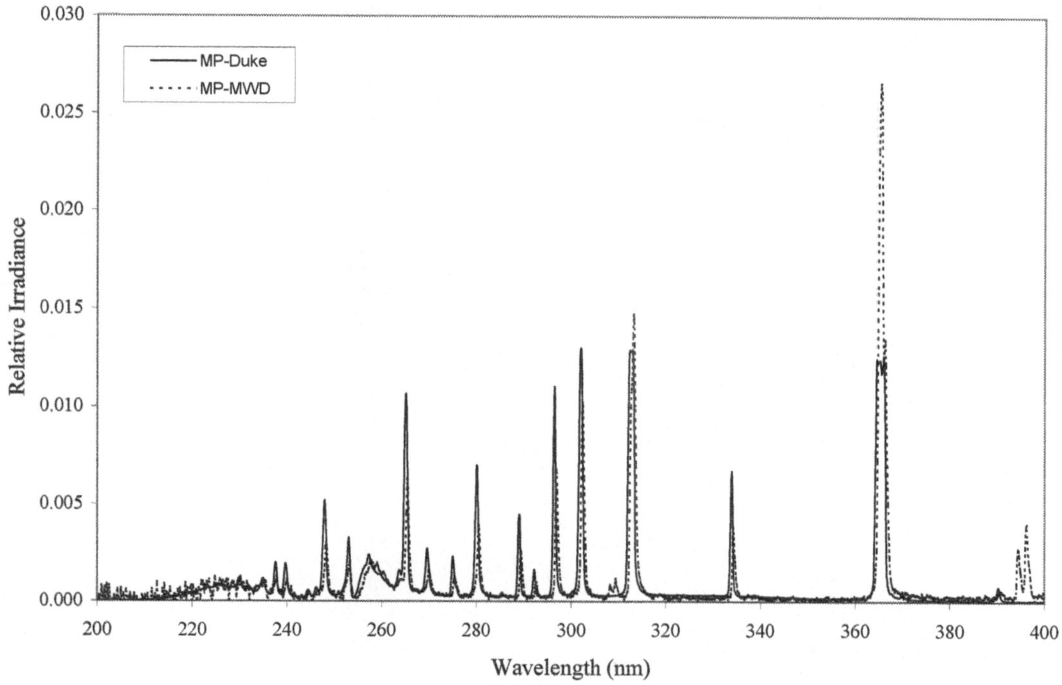

Figure 2.3 Comparison of emission spectra between two different medium-pressure ultraviolet lamps relative to total 200 – 400 nm irradiance output

Lamp Output Spectrum

The spectrum of radiation emitted from the P-UV lamp at Metropolitan was measured with the unit in the collimating mode. Figure 2.5 illustrates the germicidal UV spectra from this lamp. Figure 2.6 illustrates the total emission spectra from 200 to 900 nm. The relative contribution of the UV germicidal to the whole spectra of the P-UV lamp is strikingly small. However, within the germicidal band, there are significant outputs in the 250 to 270 nm range.

Collimating Tube Petri Factor Determination

For each lamp and collimated beam configuration, a "petri factor" was determined which characterizes the distribution of light across the petri dish. The variation in lamp irradiance across the petri dish surface was measured with a radiometer at the center of the petri dish location, and at 0.5-cm planar grid locations in both the x and y directions (25 locations). To make readings more accurate, the radiometer detector was partially blinded with an opaque paper shield with a 0.3125-inch (0.8-cm) diameter hole punched out of its center. These readings provided a map of irradiance distributions (see Figure 2.7) for each of the lamp/collimating options. By using the petri factor multiplied against a petri-dish centered radiometer reading, the geometrically averaged irradiance is then determined. The values for the average irradiance provided by each combination was found to be as follows: LP-UV lamp, 10 in. (25 cm) collimating tube = 0.80; MP-UV lamp, 20 in. (51 cm) collimating tube = 0.80; MP-UV lamp, 50 in. (127 cm) collimating tube = 0.72; and, P-UV lamp, 20 in. (51 cm) collimating tube = 0.93.

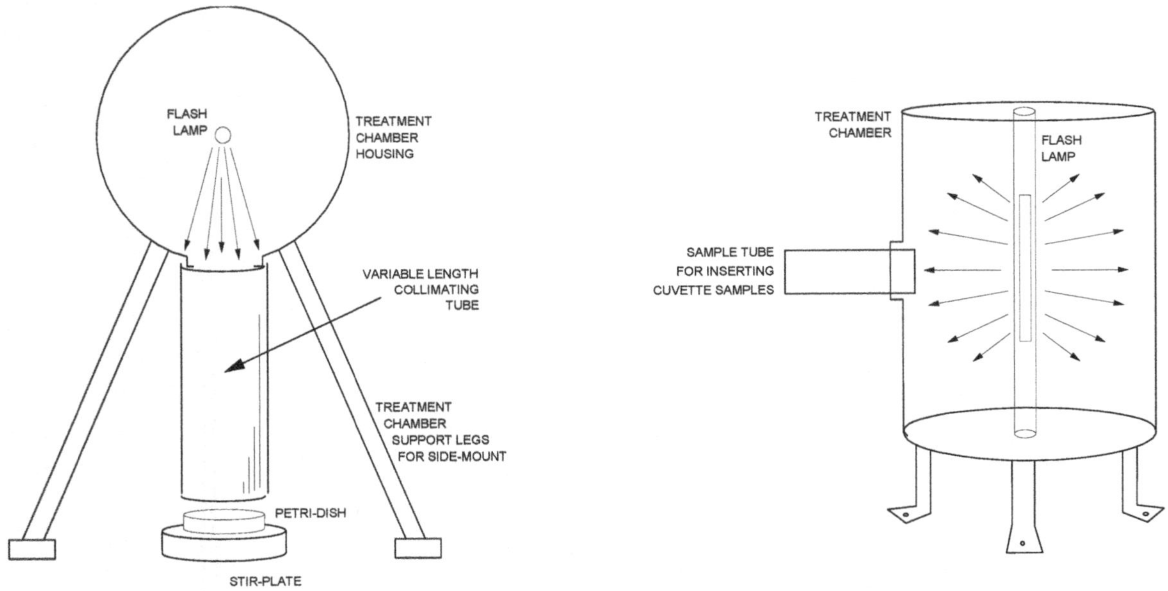

(a) collimated P-UV setup (b) horizontal P-UV setup

Figure 2.4 Illustration showing the two possible pulsed-ultraviolet sample exposure methods: (a) collimated, and (b) horizontal

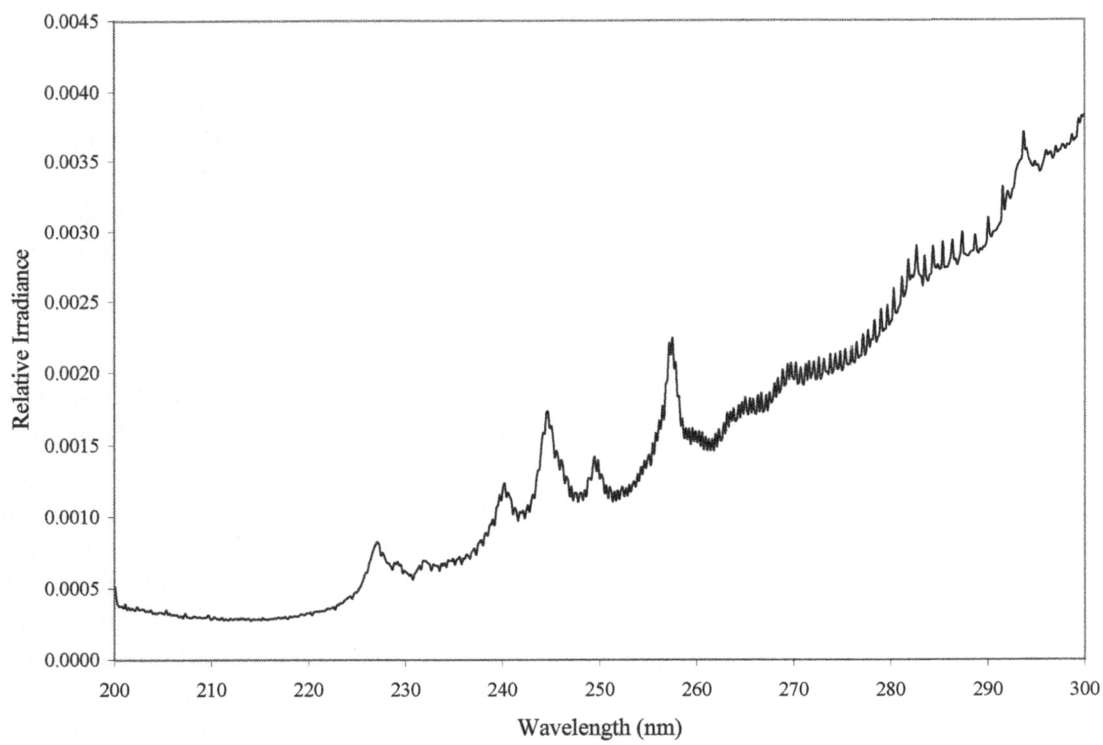

Figure 2.5 Ultraviolet, germicidal emission spectra range for a pulsed-ultraviolet, xenon arc lamp

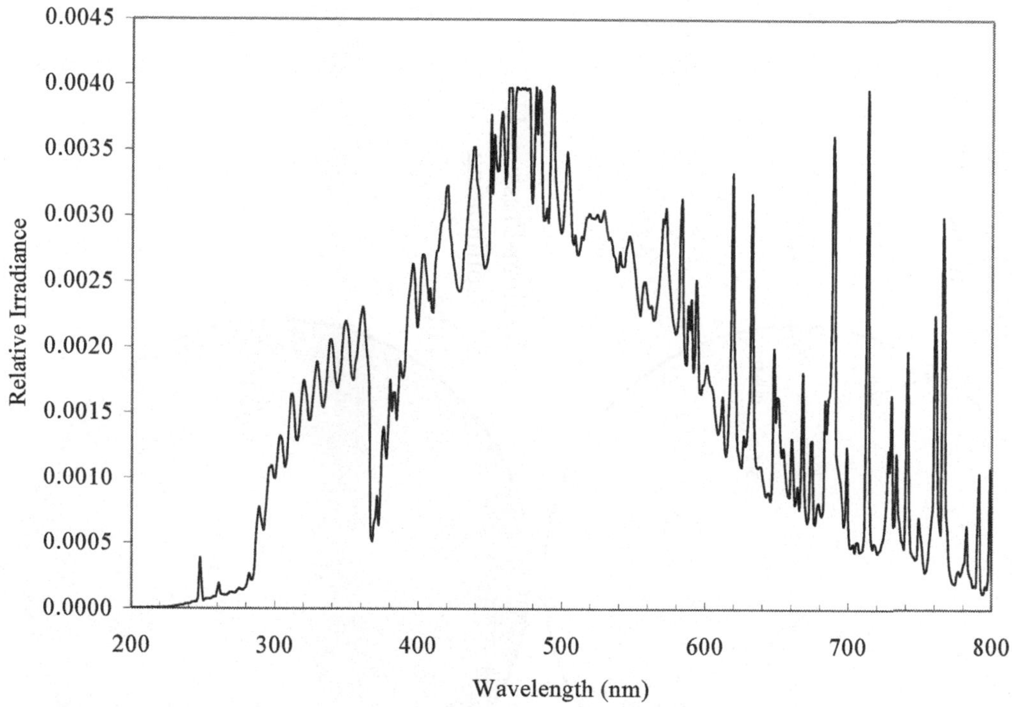

Figure 2.6 Emission spectra for a pulsed-ultraviolet, xenon arc lamp between 200 – 800 nm relative to the total output

Photoreactivation and Dark-Repair Incubators

Appendix A presents photographs of the lighted incubation box that was used for the photoreactivation experiments. The light box was fully enclosed (doors on front) with two full-spectrum daylight lamps suspended approximately 20 in. (51 cm) above the samples. There were three 1.5 cfm fans that were mounted between the two lamps on the box top. Vents were installed at the bottom sides of the light box and wooden baffles were placed in-between the samples and the vents to prevent air from rushing over the samples and causing samples to evaporate. Experiments conducted to quantify evaporation rates from samples in the light box found that from a 20 mL start volume, only 1.5 mL (7.5 percent) of sample would evaporate over a 7-day period. In the dark incubator (samples covered in aluminum foil and placed inside a laboratory cabinet), 0.8 mL of sample (4 percent) that was found to evaporate during the same time period.

DOSE MEASUREMENT METHODS

Terminology Used in This Report

An excellent discussion on terminology relating to UV disinfection was reported in the recent recommendations of the International Union of Pure and Applied Chemistry (IUPAC) Working Party on Ultraviolet Disinfection (Bolton 2000b). The terms UV "irradiance" and UV "dose" are used throughout this report and thus should be defined. The irradiance is the total

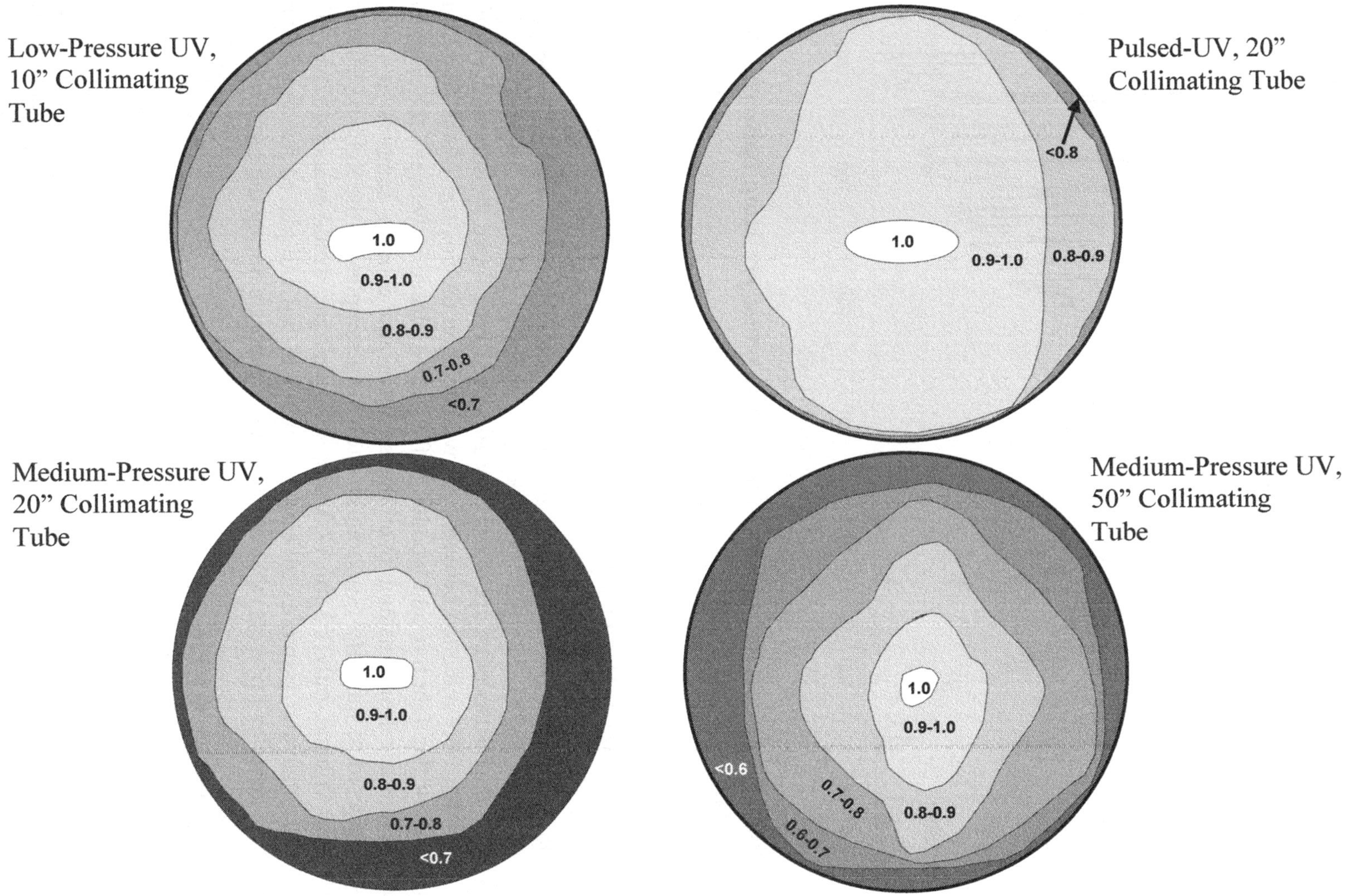

Figure 2.7 Irradiance distributions for four ultraviolet lamp and collimated tube configurations at Metropolitan

radiant power incident from all upward directions on an area. Irradiance is the appropriate term when a surface is being irradiated by UV light coming from all directions above the surface, such as in a typical bench scale set up (e.g. collimated beam). This should not be confused with the term "fluence rate" which is defined as the total radiant power incident from all directions onto an infinitesimally small sphere. Fluence rate is the appropriate term to use when UV light from many different directions, such as in a multi-lamp UV reactor, is irradiating a microorganism. Although in past literature on UV disinfection the terms "intensity" and "irradiance" have been used, it is important to realize that fluence rate is the appropriate term for UV disinfection. Irradiance and fluence rate are closely related. In a well-designed bench set-up, the fluence rate and the irradiance are the same, and the radiometer that is used with such a bench set-up measures the irradiance.

The term "UV dose" is utilized almost universally in UV disinfection literature. However, for situations in which the irradiance or the fluence rate is constant (e.g., in a collimated beam), multiplication by the exposure time (in seconds) gives the corresponding terms "radiant exposure" or "fluence", respectively. The term "fluence" is most commonly called the "UV dose". However, "dose" is a term that, in other contexts, is used to describe the total absorbed energy. In the case of microorganisms, almost all of the incident ultraviolet light passes through the organism with only a few percent being absorbed, thus the term fluence would be more appropriate since it relates to the "incident" UV energy, rather than "absorbed" UV energy. Because in the context of UV disinfection "UV dose" has become the means to describe what is actually the fluence, the term UV dose will be used through this report for historical consistency.

UV dose (H, mJ/cm^2) was calculated as the product of the UV irradiance (E, mW/cm^2) and time (t, seconds).

$$H = E \times t \qquad (2.1)$$

Because the time factor is similar for each UV system due to the use of static batch reactors, the dose measurement techniques were focused on measuring and comparing irradiance. For the P-UV lamp, the dose per pulse was used as an analog for UV irradiance provided by continuous wave UV lamps. Three conceptualizations of UV irradiance were measured as part of this research and are defined here. (1) Incident UV irradiance was a measure of the irradiance in the 200 to 300 nm range that was incident upon the surface of the sample being irradiated. (2) Average UV irradiance was a measure of the volume averaged irradiance in the 200 to 300 nm range accounting for the depth of the sample and the absorbance of the water according to Morowitz (1950). (3) Germicidal UV irradiance was the average irradiance weighted to the relative effectiveness of each wavelength in the 200 to 300 nm range, according to the absorbance spectra of DNA.

Radiometric Ultraviolet Dose Measurement

An International Light (Newburyport, MA) IL1700A Radiometer with SEL 240/G/W detector calibrated at 2 nm intervals across the UV range was used in radiometry measurement. Direct irradiance values were obtained by measuring the average irradiance across the surface area of the batch irradiation vessel (petri dish). Radiometer-based mathematical model calculations were conducted following the procedure developed by Linden and Darby (1997) and are abbreviated here in step-by-step format.

Step 1: Accounts for lamp wavelength emission and radiometer sensor responsivity, to provide a value that will translate the radiometer reading into a total incident UV irradiance value. The result of this step is a single value (specific lamp factor) representing the lamp spectra and radiometer bias.

Step 2: Utilizes the factor in Step 1 to get a value for the incident UV irradiance. Uses the radiometer reading and adds up the contribution from each line emission to arrive at the total UV irradiation incident on the sample surface.

Step 3: Determines the average irradiance in the completely mixed petri dish utilizing the absorbance spectra and depth of the water sample for each line emission of the UV lamp.

Step 4: Weights the average irradiance to a germicidal factor at each wavelength emitted by the lamp to determine the germicidal irradiance using the DNA absorbance spectra, relative to the peak absorbance wavelength as a weighting factor.

Chemical Actinometric Methods

Uridine Actinometer

0.012mM uridine (Sigma) solution in 1mM phosphate buffer (Fisher Scientific) was used as the uridine actinometer. UV dose was measured with uridine actinometry utilizing two different approaches, each giving a direct measure of incident UV dose. The first approach was the traditional method based on the principle that an absolute change in the uridine parent compound before and after irradiation related directly to the UV dose absorbed by the actinometer as detailed below.

$$H_{eff} = \frac{H_{abs}}{F} \quad (2.2)$$

$$H_{abs} = \frac{\Delta\alpha_{10} V}{\phi_{\lambda} \varepsilon A_{cs}} \times Z \quad (2.3)$$

Where

H_{eff} = effective germicidal UV dose (mJ/cm^2)
H_{abs} = UV dose absorbed by the uridine actinometer (mJ/cm^2)
F = fraction of radiation absorbed at 262 nm by the irradiated uridine
ϕ_{λ} = quantum yield of uridine degradation at wavelength of interest (moles/Einstein)
$\Delta\alpha_{10}$ = change in absorbance (cm^{-1}) coefficient (at 262 nm) of uridine due to UV exposure
ε = molar extinction coefficient (L/mol-cm at wavelength of interest)
V = volume (L) of irradiated actinometer solution
A_{cs} = cross sectional area (cm^2) of irradiated actinometer solution ($Pi*r^2$)
Z = constant used to convert Einsteins into Joules

This approach is referred to as the traditional method or the uridine change method.

The second approach was adapted from work by von Sonntag and Schuchmann (1992), Hoyer and Nick (1999), Zhang et al. (1997). The relevant equation for determining the incident irradiance from a continuous source (LP and MP-UV) is:

$$E_o = \frac{\ln(C_o/_C)}{1000 * \varepsilon\phi_\lambda t} xZ \qquad (2.4)$$

Where

E_o = incident irradiance (mW/cm^2)
C = concentration of uridine (mol/L)
C_o = initial concentration (mol/L) of uridine at time = 0
t = exposure time (t)

This approach is termed the "slope" method of determining irradiance and the dose can be calculated by multiplying this value by the exposure time.

To arrive at a value for energy per pulse, the incident energy calculated in equation 3.3 was modified to reflect the change in actinometer per pulse by replacing the time factor with a value for number of pulses. The units are mJ/cm2-pulse.

$$E_{Pulse} = \frac{\ln(C_o/_C)}{1000 * \varepsilon\phi_\lambda pulse} xZ \qquad (2.5)$$

Where

E_{pulse} = incident dose per pulse (J/cm^2-pulse)

Uridine Photohydrate Byproduct Analysis

A determination of the photoproduct for the uridine actinometer was carried out using advanced analytical techniques. Because the quantum yield of uridine is based on a photohydration reaction being the sole photodegradation product, an assurance was needed to confirm this when used for the polychromatic lamps. It was possible that the high intensity of the P-UV lamp could cause the formation of other degradation products. Therefore it was necessary to analytically determine the types of degradation products or by-products that were formed. In addition to the spectrophotometric indicator of uridine photodegradation, a semiquantitative method for uridine and the postulated major UV photoproduct, 5,6 dihydro-6 hydroxyuridine, by liquid chromatography, electrospray interface, mass spectrometry (LC/ESI/MS) was developed. Detailed methods development for photohydrate detection are presented in Appendix B.

Potassium Iodide Actinometer

As explained by Rahn (1997), and briefly described here, iodide exposed to ultraviolet light while in aqueous solution undergoes a photoionization reaction with a low quantum yield due to most aqueous electrons recombining with the iodide atoms in the solvent. Also, iodide (even as a 0.6 M solution) absorbs only out to 275 nm. Therefore, a KI-only based solution would be a poor UV actinometer. Rahn (1997) found that if iodate (IO_3^-) were added to a KI solution, the IO_3^- acted as an electron acceptor and formed a complex with the KI having two effects: (1) the absorption spectrum is shifted to the red wavelength range and (2) the quantum yield increased to approximately 0.75. Upon UV exposure, this actinometer generates triiodide ion (I_3^-) which absorbs only at 352 nm. Spectrophotometric analysis of solution absorbance at 352 nm before and after exposure to UV radiation can quantify the photon flux, and specifically, the UV dose.

Two solutions are prepared for use of KI actinometry. *Solution A* consists of a 0.010 M borax/0.10 M iodate solution made by adding 3.81 g of $Na_2B_4O_7.10H_2O$ and 21.4 g of KIO_3 to 1 L of distilled water and adjust the pH to 9.25 with 1.0 M H_2SO_4. This solution can be kept up to 30 days. *Solution B* is to be made on the day of the experiment by adding 24.9 g of granular KI to 250 mL of Solution A and stirring until dissolved. This mixture creates a solution 0.60 M in iodide.

The UV dose from the MP-UV lamp was calculated using the following equations.

$$H_{MP} = \frac{A^{S}{}_{352} - A^{Bo}{}_{352}\ V(24{,}070.70)}{(l)[1+0.02(T-20.7)]\,A} \tag{2.6}$$

Where

H_{MP} = UV dose applied from MP-UV lamp (mJ/cm^2)
$A^{S}{}_{352}$ = absorbance of the irradiated sample at 352 nm
$A^{Bo}{}_{352}$ = absorbance of the actinometer blank at 352 nm
V = the volume of sample that is irradiated [L]
l = path length of light passing through the solution [cm]
T = temperature (°C)
A = area of sample facing the light source [cm^2]
24,070.70 = constant accounting for the quantum yield, molar absorption coefficient, and conversion factor from Einsteins to mJ/cm^2 specific to the MP-UV lamp in use.

For the LP-UV lamp, the following equation was used to calculate UV dose:

$$H_{LP} = \frac{A^{S}{}_{352} - A^{Bo}{}_{352}\ V(23{,}786.40)}{(l)[1+0.02(T-20.7)]\,A} \tag{2.7}$$

Where

H_{LP} = UV dose applied from a LP-UV lamp (mJ/cm^2)

For P-UV lamp, the following equation was used to determine UV dose:

$$H_{P\text{-}UV} = \frac{A^{S}{}_{352} - A^{Bs}{}_{352} - A^{Bo}{}_{352}\ V(23{,}786.40)}{(l)[1+0.02(T-20.7)]\ n_{pulse}\,A} \tag{2.8}$$

Where

$H_{P\text{-}UV}$ = UV dose applied per pulse, not counting simmer contribution (mJ/cm^2-pulse)
$A^{Bs}{}_{352}$ = Absorbance of the simmer mode blank at 352 nm
n_{pulse} = number of UV pulses from the pulsed lamp

In research conducted by Mofidi and co-workers (2001), it was reported that P-UV, xenon lamps in simmer mode can still produce significant UV radiation without pulsing. To quantify which portion of the total UV dose is coming from the simmer mode, the following should be calculated:

$$H_{\text{simmer}} = \frac{A^{\text{Bs}}{}_{352} - A^{\text{Bo}}{}_{352}\ V(23{,}786.40)}{(l)[1+0.02(T-20.7)]\ n_{\text{pulse}}\ A} \tag{2.9}$$

Where

H_{simmer} = UV dose applied by the lamp simmer, not counting pulse contribution (mJ/cm^2)

The total applied dose given by the P-UV lamp can then be determined by the following:

$$H_{\text{P-tot}}\ [\text{mJ cm}^{-2}] = H_{\text{P-UV}} + H_{\text{simmer}} \tag{2.10}$$

Where

$H_{\text{P-tot}}$ = Total combined UV dose applied from both lamp simmer and pulses

SOURCE WATER QUALITY

Laboratory Waters

Milli-Q® Water

The laboratory Milli-Q® water (Millipore, Bedford, Mass.) was used during the photohydrate quantitation portion of the project only. This water is treated by reverse osmosis, a resin cartridge which removes trace organics and inorganics, LP-UV irradiation, and a 0.22 µm filter. Typical water quality is characterized by a low level of total dissolved solids (<10 ppb), particles (<1/mL), total organic carbon (<5 ppb), and water absorbance (0 cm^{-1}, 100 percent transmittance).

Phosphate Buffer Solution

The phosphate buffer solution (PBS) was prepared according to *Standard Method* 9050 C.1a (APHA et al. 1998). In brief, the stock PBS solution consisted of potassium dihydrogen phosphate (KH_2PO_4) diluted with organic-free laboratory water. Solution pH was adjusted to 7.2±0.5 with 1 N sodium hydroxyde. The working PBS solution made from the stock was combined with magnesium chloride solution (81.1 g $MgCl_2{\cdot}6H_2O$ in organic-free laboratory water) and autoclaved. The final solution UV absorbance ranged from 0.003 to 0.009 cm^{-1}. This corresponded to UV transmittance at 254 nm (UV_{254}) ranging from 99.4 to 98.0 percent, respectively.

Natural Waters

Ed Thomas Water Treatment Plant, City of Burlington, NC

The source water for the Ed Thomas Water Treatment Plant (WTP) is the Stoney Creek Reservoir. The process train consists of conventional alum coagulation, flocculation, sedimentation, prechlorination, filtration with anthracite/sand, pH adjustment, and chlorination. Filtered water (prior to chlorine addition) was collected in a 5-gal. carboy at one sample event, shipped to Metropolitan, and stored at 4 °C. Water from the carboy was collected when needed for testing with *B. subtilis*.

Prior to each test, various water quality parameters were analyzed. During storage and use of the water for testing (a 15-day period), the raw water pH increased from 6.2 to 7.1 units, water absorbance of 254 nm light (UV_{254}) decreased slightly from 0.11 to 0.08 cm^{-1} (an increase in water transmittance from 78 to 83 percent), water hardness remained was 37 mg/L as $CaCO_3$, and turbidity remained at 0.09 ntu. After adding *B. subtilis* to the water, UV_{254} values increased slightly, ranging between 0.10 and 0.14 cm^{-1}. This resulted in a decrease of the water transmittance which provided final values ranging from 73 to 80 percent.

Oxidation Demonstration Plant, La Verne, Calif.

Water quality data for a pretreated blend of California surface water used for many of the tests in this project is presented in Table 2.1. This blend consisted of California State Project water (SPW) and Colorado River water (CRW) conveyed to Metropolitan's La Verne, Calif. facilities. In La Verne, the blended water was treated at a 5.5-million gallons per day (MGD) treatment plant with preozonation, coagulation, flocculation, sedimentation, and dual-media (anthracite/sand) biological filtration (no residual disinfectant/oxidant added).

For all testing, water hardness, pH, and turbidity remained fairly constant while absorbance changed moderately. During the *B. subtilis* testing, filtered water UV_{254} (prior to adding *B. subtilis*) averaged 0.13 cm^{-1} (75 percent transmittance). This was somewhat higher than other microorganism testing whose average absorbance values did not rise above 0.05 cm^{-1} (transmittance remaining above 90 percent). Although the reason for this deviation in water quality was unknown, it was incorporated into the standard UV dose measurement procedure presented in this report so that delivered UV dose would remain the same for all exposures. Except during *B. subtilis* testing, adding microorganism suspensions to the filtered surface water produced very small increase in the water's absorbance value. For example, the UV_{254} for the filtered water used for MS-2 testing averaged 0.05 cm^{-1} (90 percent transmittance) before and after adding MS-2. For the *Cryptosporidium* tests, average water absorbance increased by 0.01 cm^{-1} (transmittance decreased by 1 percentage point). Although changes were typically small, they were also compensated for during measurement of delivered UV dose.

Sweeney Water Treatment Plant, City of Wilmington, NC

The source water for the Sweeney WTP is the Cape Fear River. A pumping station conveys the water through 25 mi. of pipeline to Wilmington. The process train consists of

Table 2.1
Average ± standard deviation of water quality parameters measured of California State Project water/Colorado River water blend used in microorganism testing

Test Organism	Absorbance 254 nm (cm^{-1})	Hardness (mg/L as $CaCO_3$)	pH (units)	Transmittance 254 nm (percent)	Turbidity (ntu)
*Bacillus subtilis**	0.13 ± 0.05	227 ± 5	7.97 ± 0.09	75 ± 9.3	0.08 ± 0.02
Bacillus subtilis†	0.18 ± 0.02			66 ± 3.2	
MS-2 coliphage*	0.05 ± 0.00	208 ± 6	8.03 ± 0.10	90 ± 0.7	0.07 ± 0.01
MS-2 coliphage†	0.05 ± 0.00			90 ± 0.3	
Heterotrophic bacteria‡	0.03 ± 0.01	204 ± 15	7.88 ± 0.13	93 ± 1.9	0.11 ± 0.07
Cryptosporidium (all tests)*	0.03 ± 0.00	230 ± 14	8.07 ± 0.07	93 ± 0.7	0.07 ± 0.02
Cryptosporidium†	0.04 ± 0.00			92 ± 0.7	
*Cryptosporidium***	0.03 ± 0.00			93 ± 0.7	

* Before adding microorganisms in suspension
† After microorganism suspension (absorbance values after suspending *Cryptosporidium* were measured in limited experiments)
‡ Organisms were naturally occurring
** Data for water prior to suspending *Cryptosporidium* for the limited number of tests where absorbance was measured both before and after suspending *Cryptosporidium*

pre-ozonation of raw source water, followed by either conventional coagulation-sedimentation process or coagulation-upflow clarification process, all waters are then ozonated again prior to filtration with GAC-capped, sand and gravel filters operated in biofiltration mode.

Similar to the water from Burlington, this water was collected after filtration and sent to Metropolitan for *B. subtilis* testing in one 5-gal. carboy and stored at 4 °C. Water pH increased over a 1-month test period from 6.1 to 6.4 units, UV_{254} decreased from 0.17 to 0.05 cm^{-1} (a transmittance increase from 68 to 89 percent), hardness decreased from 26 to 23 mg/L as $CaCO_3$ (change possibly due to analytical error), and turbidity increased from 0.11 to 0.12 ntu. *B. subtilis* addition caused UV_{254} to increase slightly, ranging between 0.05 and 0.22 (transmittance ranging between 61 and 89 percent).

ANALYTICAL METHODS

Standard Methods

Several analytes were measured by *Standard Methods* (APHA 1998). These include alkalinity (titration, 2320 B), hardness (titration, 2340 C), heterotrophic bacteria (see text below), total organic carbon (UV/persulfate oxidation, 5310 C), and ultraviolet light absorbance at 254 nm (UV/vis spectrophotometer without pre-filtering samples, 5910).

Bacillus subtilis

Stock **Bacillus subtilis** *Preparation*

On aged, dried nutrient agar or trypan blue agar plates, an aliquot of *B. subtilis* (ATCC 6633) was grown as a pure culture for 2-3 days at 35 °C. Then, the culture was incubated for 3-4 days at room temperature to starve the organisms (spores form when the nutrients are exhausted). The organisms were then scraped from the plate and resuspended in sterile distilled water. The stock solution was then heated at 82 °C for 15 min to kill vegetative cells, immediately put on ice, and kept at 4 °C.

Experiment Samples

Positive controls consisted of *B. subtilis* stock solution spiked into sterile water and phosphate buffer solution (PBS). On average, numbers of *B. subtilis* for LP-, MP-, and P-UV tests were 2.6×10^6, 2.7×10^6, and 2.7×10^6, respectively. There was a very low standard deviation of 4×10^5 for each of the data sets. Negative controls consisted of sterile PBS. Sterility of PBS was assessed with each run by filtering an aliquot of PBS and incubating the membrane at 35 °C for 24 h. Media sterility was assessed during each experiment by incubating 2 plates with the rest of the samples at 35 °C for 24 h.

The assay procedures for determining the number of live spores included aseptically filtering controls and treated samples through a sterile, 0.45-µm, 47-mm membrane (Gelman Sciences #66068). This was done using autoclaved, sterile, glass filter holders and PBS. Plates were incubated at 35 °C for 22 ± 2 h. Spores were then counted. Results were reported as colony forming units per millileter (CFU/mL).

Trypan Blue Agar Media Preparation

On a heated stir plate, the following reagents were mixed:

- 15.0 g agar (Difco)
- 3.0 g dessicated beef extract
- 5.0 g peptone
- 0.03 g trypan blue
- Complete to 1000 mL with Milli-Q® water

The mixture was autoclaved for 15 min., cooled to 45 °C using a water bath, then poured into 60x15 mm petri plates. For quality control (QC), one plate was incubated at 35 °C each week. If growth was detected on the QC plate, the mixture was discarded and reformulated.

Male-Specific Coliphage MS-2

The methods used in preparing and quantifying the MS-2 coliphage were developed based on previous research (Adams 1959, Mofidi et al. 1998). MS-2 coliphage was handled in either glass or polypropylene containers only due to its possible affinity to adsorb to other materials. MS-2 solutions were kept at 4 °C (samples were immediately placed on ice when taken out of the refrigerator) unless otherwise specified.

***Escherichia coli* Preparation**

In a sterile test tube, 10 mL of tryptone broth was inoculated with one full loop (10 μL) of frozen culture of *E. coli* Famp and incubated overnight at 35 ± 1°C. Before collecting the *E. coli* aliquot, the solution was vortexed to permit *E. coli* reconstitution and growth after storage at -20°C. After incubation, 0.2 mL of the overnight culture was transferred into 50 mL of tryptone broth in a 250 mL erlenmeyer flask (or 0.4 mL of culture into 100 mL broth, 250 mL flask). This was incubated for 4-6 h at 35 ± 1°C under gentle agitation (125 rpm in a shaker table). The goal of this step was to allow maximum pili expression, the site at which MS-2 coliphage invades *E. coli*.

***Escherichia coli* Inoculation With MS-2**

Prior to inoculation, the following items were prepared: labeled bottom agar petri plates, tryptone top agar tubes in water bath at 45 ± 1 °C, *E. coli* 4-h culture, MS-2 samples and sample dilutions, and 1- and 10-mL pipets. No replicates were required for sample dilution. Each sample and dilution were run in triplicate petri plates. Each MS-2 sample was diluted by adding either 0.4 mL into 3.6 mL of saline-calcium solution or 0.3 mL into 2.7 mL. MS-2 samples and dilutions were kept at 4 °C due to the temperature sensitivity of MS-2.

Working in the water bath and using a 5-mL sterile pipet, 1.0 mL of *E. coli* 4-h culture was added to no more than 20 top agar tubes. A 1.0-mL aliquot of MS-2 sample or diluted sample was then added to the tubes. Immediately, tubes were removed from the water bath and mixed by rolling rapidly in-between the palms of the analyst's hands for a few seconds while avoiding the formation of bubbles. The mixture was then dispensed over a bottom agar petri plate. *E.coli* remained at 45 °C in tryptone top agar for less than 15 min and MS-2 remained for less than 1 min. Plates were rotated to spread solution evenly over the bottom agar, allowed to solidify, then inverted and incubated at 35 ± 1 °C for 22 ± 2 h.

Plaques formed on each petri plate were then counted (dilutions of sample were made to target between 20-100 plaques). Average plaque counts of the 3 plate replicates were multiplied by the reciprocal of the dilution and divided by the MS-2 sample volume added. Results are reported in plaque forming units per mL (PFU/mL).

Escherichia coli Stock Solution

The pure culture of *E. coli* Famp (ATCC 15597) was resuspended in tryptone broth and vortexed. After splitting in cryovials, 10 percent sterile glycerol was added. Pure *E. coli* culture was either kept at -20 °C for regular utilization and short-term storage or -80 °C for long-term storage.

MS-2 Stock Solution

An adequate amount of petri plates (as described above under the section *E. coli* Inoculation with MS-2) were inoculated using an MS-2 (ATCC 15597-B1) stock solution of 1,000 to 2,000 phage/0.1 mL. After a 24-h incubation at 35 °C, 5 mL of saline-calcium solution was added to each plate, allowed to remain for 30 min, and then the coliphage was recuperated. After tilting the plate, approximately 4 mL of liquid containing phage, *E. coli*, and agar was collected from each plate. This was then transferred into sterile 250-mL polypropylene centrifuge flasks. Centrifuging was done for 15 min at 10,000*g* and 4 °C. After centrifugation, the upper phage-rich layer of the flask was collected and filter sterilized with a pre-treated (with tryptone broth in order to avoid phage adsorption to the membrane), disposable syringe filter (0.2-µm cellulose acetate membrane) to remove *E. coli*. This collected phage was then split into sterile glass flasks or bottles and stored at 4 °C. The MS-2 stock solution was monitored weekly by measuring its concentration as described above (inoculation of *E. coli* host with MS-2 sample), and prepare new phage solution when concentration starts to decrease, or when the solution is older than 6 weeks.

Purification of MS-2 Stock Solution

The MS-2 stock solution was purified to remove light-absorbing compounds which might interfere with disinfection testing. To accomplish this, chromatography columns (TE Super Select-D, G-50 Spin Columns; 5'→3', Inc., Boulder, Colo.) were used to purify plasmid and duplex DNA greater than 72 bp. Purification included inverting the column at least 3 times to resuspend the gel, leaving no resin in top closure. After removing the top (large) closure followed by the bottom (small) closure, the column was placed in the collection tube. The columns were then centrifuged for 3 min at 1,100*g* at 2,300 rpm and 4 °C with a swinging bucket or horizontal rotor. Next, columns were removed and the collection tubes were emptied.

Empty columns were placed back into the sample collection tube and washed with 250 µL of sterile saline-calcium solution. The solution was added to the column and then they were centrifuged for 3 min at 1,100*g* (2,300 rpm) and 4 °C with a swinging bucket or horizontal rotor. Washing was repeated 4 more times (total of 5 washes) and wash solution was discarded.

After removing the column, it was placed in a new collection tube and 250 μL of MS-2 stock solution was added. This was centrifuged for 5 min at 1,100*g* (2,300 rpm) and 4 °C with a swinging bucket or horizontal rotor. Columns were discarded and phage was recuperated in the collection tube. If more than one column was used, all purified phage was combined.

Experiment Sample Handling

UV-exposed samples and experiment controls were analyzed during each experiment. All samples and controls were kept on ice at 4 °C during and after experiments. After exposing MS-2 samples to UV light they were transferred into sterile 15-mL VWR falcon tubes. The phage stock solution positive control was made by pipetting an aliquot of MS-2 stock solution into one sterile collection tube. Negative controls consisted of sterile saline-calcium solution and the MS-2's support water matrix. Other experiment controls included an *E. coli* control (a petri plate prepared with top agar and *E. coli*, without the addition of MS-2) and a media control (bottom agar plate left in the incubator after *E. coli* inoculation to examine for possible agar contamination).

Miscellaneous Solutions Needed for MS-2 Assay

Saline-calcium buffer solution. Added together were 8.5 g NaCl and 0.22 g $CaCl_2$ into 1,000 mL Milli-Q® water. The solution was sterilized by autoclaving and kept with a holding time of 6 months.

Tryptone broth. In a 250-mL Erlenmeyer flask (or larger), the following items were combined: 1.0 g tryptone (Difco #0123 or equivalent); 0.1 g yeast extract (Difco #0127 or equivalent); 0.1 g glucose; 0.8 g NaCl; and, 0.022 g $CaCl_2$. The solution was completed to 100 mL with Milli-Q® water and dissolved by autoclaving. The broth was kept at 4 °C and had a holding time of 6 months.

Tryptone bottom agar (petri plates). In a 250-mL Erlenmeyer flask (or larger), the following items were combined: 1.0 g tryptone (Difco #0123 or equivalent); 0.1 g yeast extract (Difco #0127 or equivalent); 0.1 g glucose; 0.8 g NaCl; 0.022 g $CaCl_2$; and, 1.5 g Bacto-agar (Difco #0140 or equivalent). The solution was completed to 100 mL with Milli-Q® water and dissolved by autoclaving. After autoclaving, the solution was dispensed in 10-15 mL aliquots into sterile 100 x 15 mm petri plates (use petri plates without second inner rings which can interfere with plaque readings). The solution was allowed to harden after storage at 4 °C. The bottom agar was then warmed to room temperature 1 h before use. Holding time for the agar was 4 weeks.

Tryptone top agar (petri plates). In a 250-mL Erlenmeyer flask (or larger), the following items were combined: 1.0 g tryptone (Difco #0123 or equivalent); 0.1 g yeast extract (Difco #0127 or equivalent); 0.1 g glucose; 0.8 g NaCl; 0.022 g $CaCl_2$; and, 1.0 g Bacto-agar (Difco #0140 or equivalent). The solution was completed to 100 mL with Milli-Q® water and dissolved by autoclaving. After autoclaving, the solution was dispensed in 3 mL aliquots into sterile, disposable, 16 × 100 mm glass culture tubes. The solution was allowed to harden after storage at 4 °C. Before use, the agar was autoclaved and placed in a 45 ± 1 °C. Holding time for the agar was 3 months.

Heterotrophic Bacteria

Heterotrophic bacteria studied in this project occurred naturally in the ozonated, biologically-filtered California surface water blend from Metropolitan's demonstration plant in La Verne, Calif. Approximately 1 x 10^4 cfu/mL of HPC bacteria occurred naturally in the biological filter effluent of the plant. Experiments requiring HPC analysis included membrane filtration and incubation on R2A media at 28 °C for 7 days, according to *Standard Methods* (APHA et al. 1998). Plating was conducted in triplicate.

Cryptosporidium parvum

The assay procedure for *C. parvum* was similar to that reported by Mofidi and co-workers (2001). Calves at the University of Arizona Department of Veterinary Science (Tucson, Ariz.) were infected with *C. parvum* genotype 2 Iowa isolate originally from the National Animal Disease Center (Ames, Iowa). Harvested oocysts were cleaned by sequential centrifugation through sucrose and cesium chloride, placed in antibiotic solution (100 ug/mL penicillin, 100 ug/mL streptomycin, and 100 ug/mL gentamicin) containing 0.01 percent polyoxyethylenesorbitan monolaurate (Tween 20, Sigma Chemical Co., St. Louis, Mo.) and shipped to Metropolitan's Water Quality Laboratory in La Verne, Calif. Oocysts were stored for less than 2 weeks at 4 °C until used in experiments.

Before UV tests, oocysts were purified by immunomagnetic separation with paramagnetic beads (Dynabeads®, Dynal Inc., Lake Success, N.Y.) and then counted by hemocytometer (Bright-line, Hausser Scientific, Horsham, Pa.). Excystation rates were always >90 percent. Suspensions of *C. parvum* oocysts contained a known number of oocysts in pre-treated, oxidant-free water collected from the biological filter effluent of Metropolitan's demonstration plant. As a result of oocyst purification, subsequent suspension in the natural water did not significantly change the water's UV absorbance characteristics (see water quality information reported in Table 2.1). On average, oocyst addition caused water absorbance to increase less than 0.01 cm^{-1} (less than 1 percentage point of UV transmittance).

Samples exposed to UV light (and experiment controls) were processed immediately. Processing included centrifugation at 1,050x*g* (15 min at 19 °C) to produce a 1-mL sample, serial dilutions of samples and controls, and inoculation onto cell monolayers (infectivity was determined by a cell culture method described by Rochelle et al. [1997]). Confluent monolayers of human ileocecal carcinoma cells (HCT-8 and ATCC CCL 244, American Type Culture Collection, Rockville, Md.) were grown in single-well slides (Lab-Tek, Nalge Nunc International, Rochester, N.Y.) over two to three days then infected with *C. parvum* oocysts and incubated at 37 °C for 48 h in 5 percent carbon dioxide. Total RNA was extracted with a purification kit (S.N.A.P. kit, Invitrogen Corp., Carlsbad, Calif.) and the messenger RNA fraction was purified using oligo (dT) cellulose (Micro-Fast Track kit, Invitrogen Corp., Carlsbad, Calif.). Reverse-transcriptase, polymerase chain reaction (RT-PCR) was used to detect *C. parvum* heat-shock protein 70 gene transcripts using primers described by Rochelle et al. (1997). After gel electrophoresis, PCR products were confirmed by hybridization with a *C. parvum*-specific oligonucleotide probe.

To quantify the degree of *C. parvum* infection, gel images were digitized on a flatbed scanner (Epson ES-1200C, Sieko Epson Corp., Nagano-ken, Japan). Calibration curves were constructed by plotting log concentrations of oocysts which were confirmed by staining procedures conducted by immuno fluorescence assay (IFA). The IFA staining procedure used was according to USEPA (1999). The intensities of the IFA-confirmed calibration curves from the scanned gel images were read by a commercially available gel-quantitation software (Pro-Quanta®, DNA ProScan Inc., Nashville, Tenn.). Then, intensities for sample gels were compared to the IFA-confirmed calibration curves to determine the number of infectious oocysts remaining after exposure to UV light.

Experiments included negative and positive controls. Negative controls were comprised of non-infected slides, heat-inactivated (70 °C for 10 min) oocysts on the cell monolayer, and presumedly oocyst-free, biofilter-effluent water on the cell monolayer. RT and PCR negative controls (molecular-grade water in the RT or PCR mixes) were also processed. If any negative controls showed signals for the presence of infective oocysts, the experiment was discarded. Positive controls included (1) amplification of a 600-bp amplicon from the human GAPDH gene after RNA extraction, (2) previously extracted DNA of *C. parvum*, and (3) *C. parvum* oocysts first placed in the petri dishes and inserted under the collimating tubes of the UV units but not exposed to UV light then inoculated onto cell monolayers.

CHAPTER 3
DOSE MEASUREMENT FOR POLYCHROMATIC UV SYSTEMS

INTRODUCTION

UV irradiance was measured for three different UV lamp types on the bench scale using three different measurement techniques: radiometry, chemical actinometry, and biodosimetry. The goal of this work was to develop and evaluate techniques for measuring germicidal UV dose from each source such that these methods could subsequently be used to compare dose-response relationships for various microorganisms across the different UV technologies. Because the irradiance is unique to each UV lamp type due to differences in lamp emission output, this parameter was followed and evaluated for each UV technology independent of time. With each irradiance measurement technique used, values for three measures of irradiance were produced for comparison including incident irradiance, average irradiance, and germicidal irradiance (described previously in Chapter 2). Each measure of UV irradiance was mathematically calculated from a direct reading of irradiance produced from each measurement technique. The direct, incident, average, and germicidal irradiance values for each lamp type are reported.

RADIOMETER APPROACH TO IRRADIANCE MEASUREMENT

Radiometer readings were taken using an International Light (Newburyport, MA) IL1700A Radiometer with SEL 240/G/W detector calibrated across the UV range at 2 nm intervals. Direct irradiance values were obtained by measuring the petri factor corrected irradiance across the surface area of the batch irradiation vessel (petri dish). Direct dose was obtained with direct irradiance multiplied by the exposure time. Values for incident, average, and germicidal irradiance were calculated using the methods described in the Chapter 2.

ACTINOMETRY APPROACH TO IRRADIANCE MEASUREMENT

Two different actinometers were utilized during the experiments on irradiance measurement. These included uridine and potassium iodide/iodate (KI). Uridine and KI actinometers were used principally as they appeared to have the most promise for routine use given that they were not sensitive to visible light. For the uridine and KI actinometers, measurements were made under each UV lamp type for direct irradiance using the techniques and equations described in the Chapter 2. To calculate values for incident, average, and germicidal UV irradiance, mathematical corrections were developed for each actinometer as described below.

Mathematical Analysis for Actinometers

Measurement of direct irradiance from the chemical actinometers (CA) were incorporated into mathematical model analyses as illustrated in the flow charts in Figures 3.1 to 3.3. Figure 3.1 illustrates the computation of the incident irradiance (within the 200 to 300 nm range) for the KI actinometer. For the computation of the incident irradiance with the uridine actinometer, the steps in the dashed line box in Figure 3.1 were substituted with the steps shown

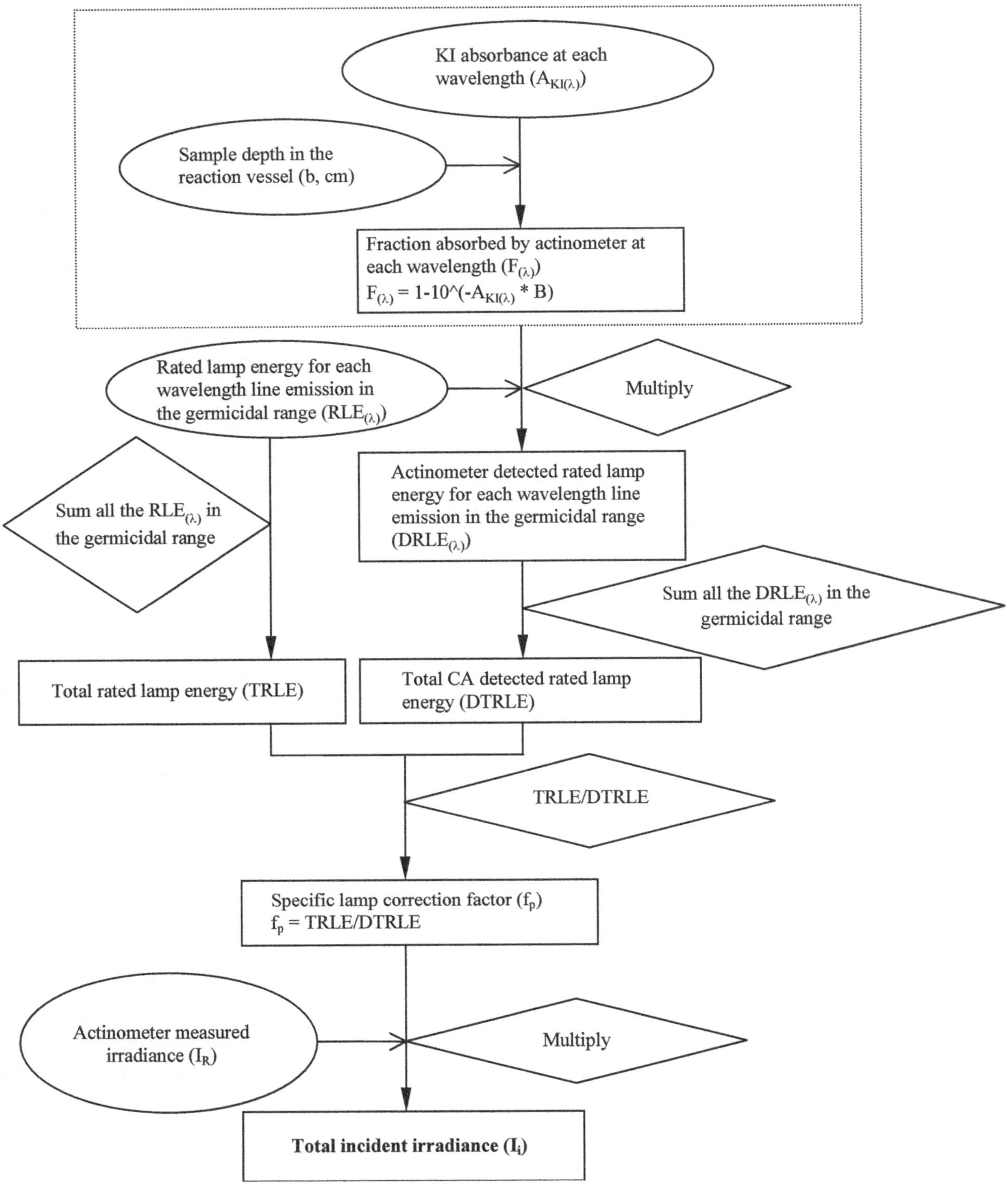

Figure 3.1 Computation of total incident irradiance from the KI actinometer combined with a mathematical analysis method

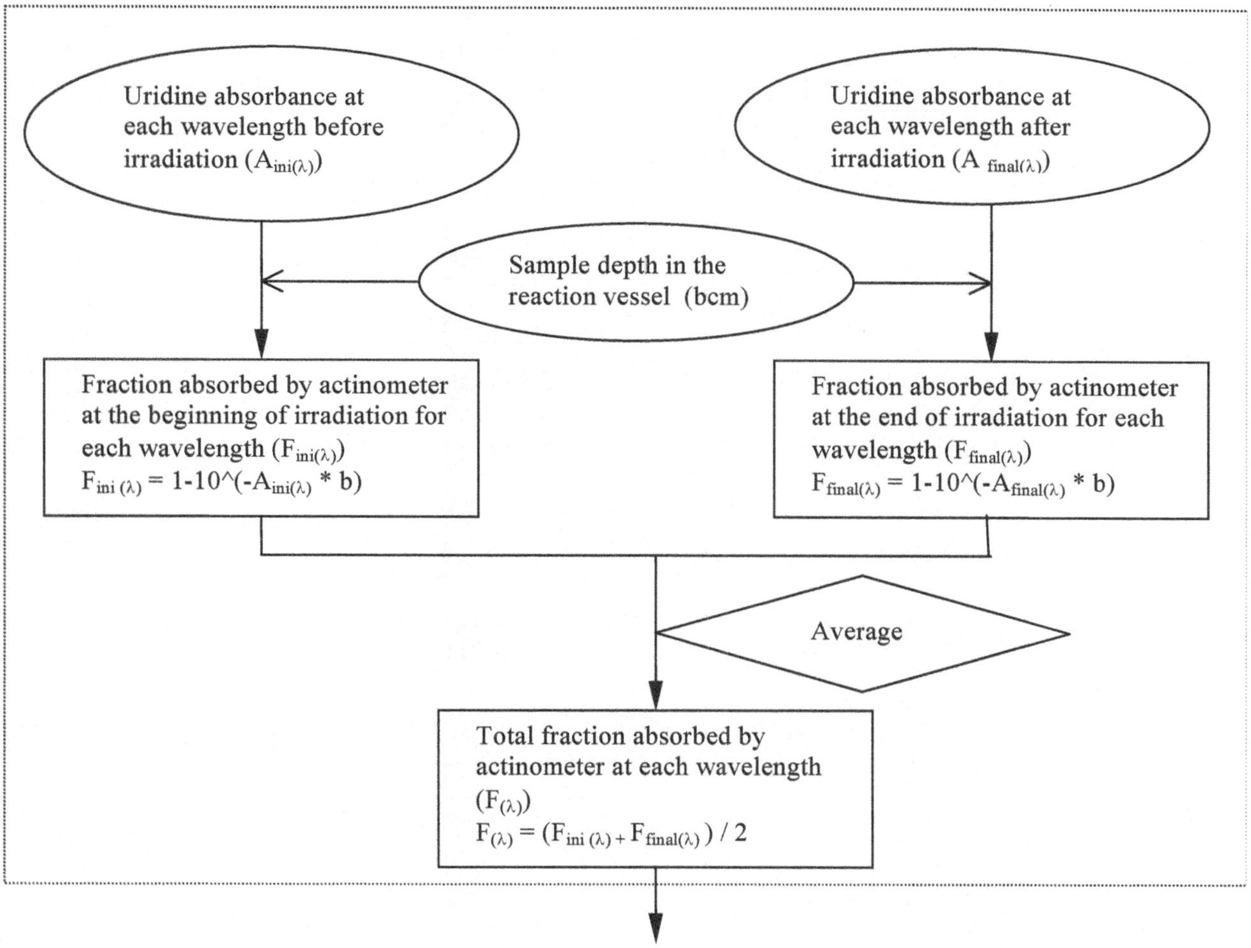

Figure 3.2 Computation of the total fraction of irradiance absorbed by the uridine actinometer

in Figure 3.2. The calculations for the actinometric average irradiance and the germicidal irradiance were computed following the steps in the flowchart presented in Figure 3.3.

In the flowcharts, the oval boxes contain values that are measured and used as inputs. The rectangular boxes show the calculated values. Diamond boxes are the operations. All the absorbance values were measured using a Cary 100 Bio UV-Vis spectrophotometer from Varian (Palo Alto, CA). The rated lamp energy for each wavelength line emission in the germicidal range ($RLE_{(\lambda)}$) was obtained using a S2000 spectrometer from Ocean Optics, Inc (Dunedin, FL). The actinometer-measured irradiance was obtained by calculating the dose values according to the methods described previously for each actinometer and dividing the dose by the exposure time.

Similar to the radiometer-based mathematical model, the chemical actinometer-based mathematical analysis method was developed to account for the impacts of (1) difference in fraction of energy absorbed by the chemical actinometer at different emission wavelength from polychromatic source, (2) changes of fraction absorbed by actinometer during exposure, (3) the water absorbance on the irradiance, and (4) the fact that not all wavelengths emission within germicidal range are equally germicidal effective are accounted for the in model. As shown in

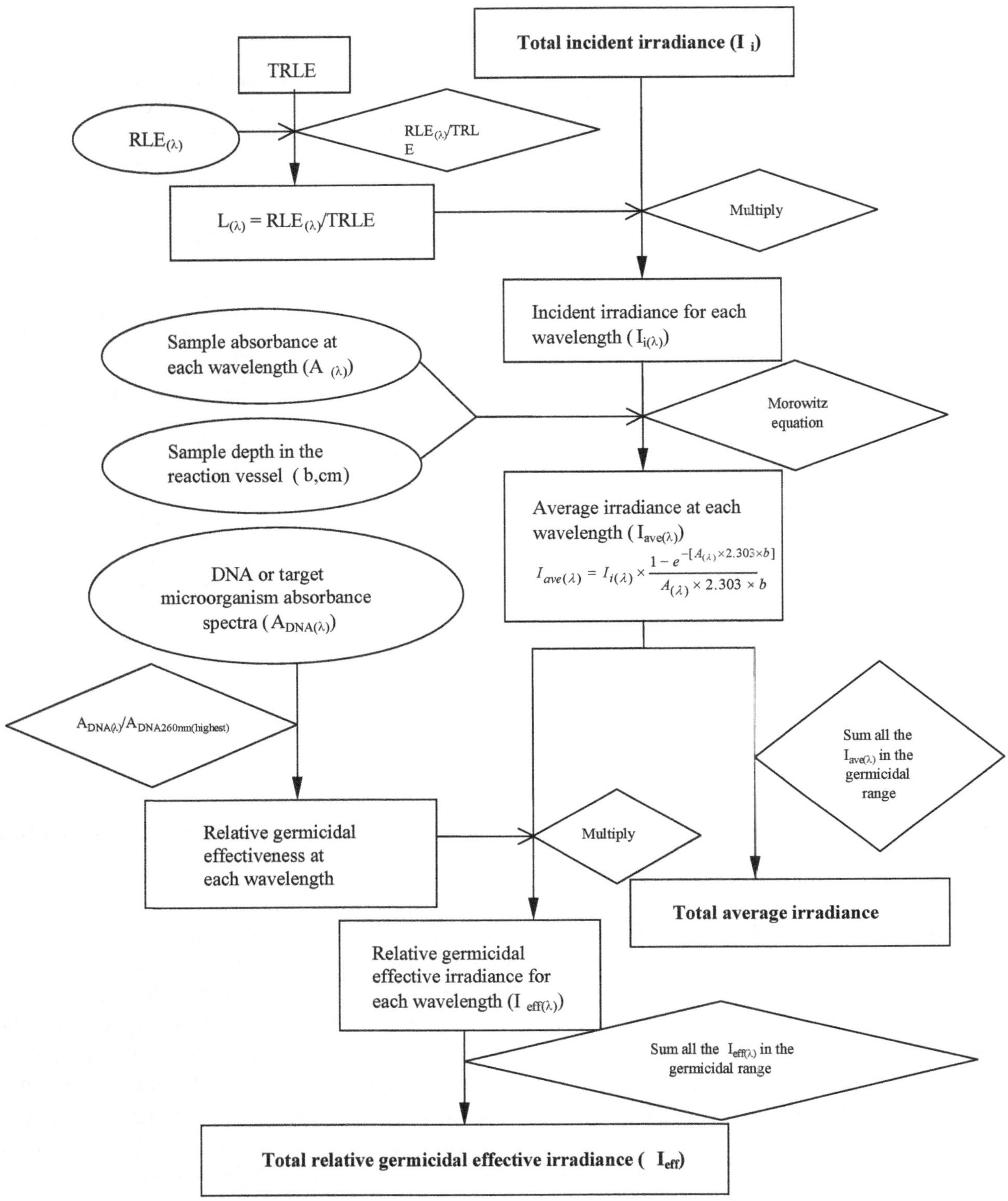

Figure 3.3 Computation of average irradiance and relative germicidal effective irradiance from chemical actinometer combined with mathematical analysis method

the flowcharts, in order to obtain the germicidal effective irradiance, the specific UV lamp emission spectra was combined with the fraction of light absorbed by the actinometer at each wavelength resulting in a lamp correction factor to modify the actinometer measured irradiance. The result is the incident irradiance in the 200-300 nm range. Water sample absorbance spectra were then incorporated into this calculation to reflect the impact of water absorbance and obtain an average irradiance of UV. The average irradiance was then weighted by the DNA absorbance spectra to yield the germicidal irradiance.

The difference in computation procedure for fraction of light absorbed by actinometer at each wavelength ($F_{(\lambda)}$) for uridine and KI actinometer accounted for the fact that uridine absorbance in 200 to 300nm range changed during the irradiation, thus the $F_{(\lambda)}$ values changed linearly with the absorbance. The absorbance and UV fraction absorbed by KI actinometer did not change significantly during the irradiation, nor did the absorbance profile.

COMPARISON OF RADIOMETER AND ACTINOMETER IRRADIANCE MEASUREMENTS

Results from irradiance measurements using actinometry and radiometry on LP-UV, MP-UV and P-UV lamps are tabulated in Table 3.1. Graphical representation of extended actinometry/radiometry testing for all three lamps is presented in Appendix D. Each actinometer measurement of irradiance is paired with a radiometer measurement taken at the same time. Data for the direct measurement (no mathematical correction), calculated incident irradiance, and germicidal incident irradiance are presented. Because a real water sample was not used for these tests, a value for the average irradiance was not calculated.

Table 3.1
Results from mathematical modeling of incident and germicidal ultraviolet irradiance using radiometer and actinometer raw data

Lamp Type	Irradiance (mW/cm^2)	Radiometer	KI	Radiometer	Uridine
	Direct measure*	0.14	0.13±0.001	0.16	0.14±0.020
LP	Incident†	0.15	0.13±0.010	0.17	0.13±0.030
	Germicidal‡	0.12	0.11±0.001	0.13	0.10±0.020
	Direct measure	0.24	0.24±0.010	0.24	0.18±0.010
MP	Incident	0.35	0.32±0.010	0.35	0.34±0.060
	Germicidal	0.15	0.13±0.003	0.15	0.14±0.020
	Direct measure	0.05	0.04±0.002	0.05	0.06±0.011
P-UV**	Incident	0.69	0.35±0.014	0.69	2.24±0.710
	Germicidal	0.02	0.01±0.001	0.02	0.07±0.020

* Incident by meter: incident irradiance obtained from radiometer reading or regular actinometer irradiance calculation method

† Incident by math model: incident dose after mathematical analysis

‡ Germicidal: relative effective germicidal irradiance obtained from weighted DNA response R_{DNA}

** Unit for P-UV is dose/pulse, not irradiance.

Low-Pressure Ultraviolet Irradiance Determination

Comparing direct irradiance calculated by radiometer and actinometer and the incident irradiance calculated after mathematical analysis, LP data do not differ greatly between radiometer or actinometers. Because approximately 90 percent of the LP lamp emission is at 253.7 nm, the radiometer should be able to accurately record all the emission within this range as long as the correct calibration factor (254 nm) for the radiometer is used. For the KI and uridine actinometers, the direct and incident values are similar. The germicidal weighting for the LP-UV values assumes that 254 nm is not the peak germicidal effective wavelength, thus its germicidal weighting factor is less than 1. In practice, all irradiation at 254 nm is considered germicidal and reported un-weighted. Overall, the radiometer and actinometer methods were able to produce similar measures of UV irradiance from the LP-UV lamp.

Medium-Pressure Ultraviolet Irradiance Determination

Because of the polychromatic nature of the MP-UV lamp, a mathematical correction is necessary to represent the incident, average, and germicidal irradiance values. After mathematical analyses, incident irradiance predicted by the radiometer and actinometer are similar to each other (within 10 percent for KI, 3 percent for uridine), which suggests the true MP incident irradiance on the water surface. These results support the need for mathematical analysis toward an accurate MP incident irradiance prediction. In addition, the results also help in understanding the germicidal irradiance measurement. It appears that the "germicidal" filter, which is weighted to provide a direct measure of the germicidal irradiance (incident by meter), does not provide the same irradiance as the mathematically corrected germicidal irradiance (based on the DNA absorbance spectra). This may be understood as an artifact of the relative weighting of the germicidal correction factor used as compared to the weighting on the "G" filter. This discrepancy was tested by weighting the "incident by math model" value by the actual germicidal weighting of the G filter to check if it equaled the "incident by meter" reading. This calculation yielded results identical to the "incident by meter reading" as expected because it was this same sensor correction factor that was used in derivation of the "incident by math model" irradiance. Therefore, it is evident that the germicidal weighting factor chosen greatly affects the calculation of germicidal irradiance. The germicidal weighting of the two methods are compared in Figure 3.4. Data collected with the G filter does not include any wavelengths below 220 nm while the DNA absorbance weighting does. In addition, the G filter has greater weight placed on the higher wavelengths. Therefore, for incident irradiance measures the germicidal irradiance value may differ as a function of the specific weighting factor used. In previous work (Linden et al. 2001) the germicidal wavelength effectiveness for *C. parvum* matched the germicidal weighting of the G filter very well. The G filter weighting also match the data generated by Gates (1929) for *E. coli* action spectra. These findings are significant in that they suggest the DNA absorbance spectra may <u>not</u> be an ideal weighting factor for estimating germicidal UV irradiance. One factor omitted here is the absorbance of the solution, which would neutralize the inclusion of low wavelengths in the germicidal irradiance calculation and may be an equalizing factor.

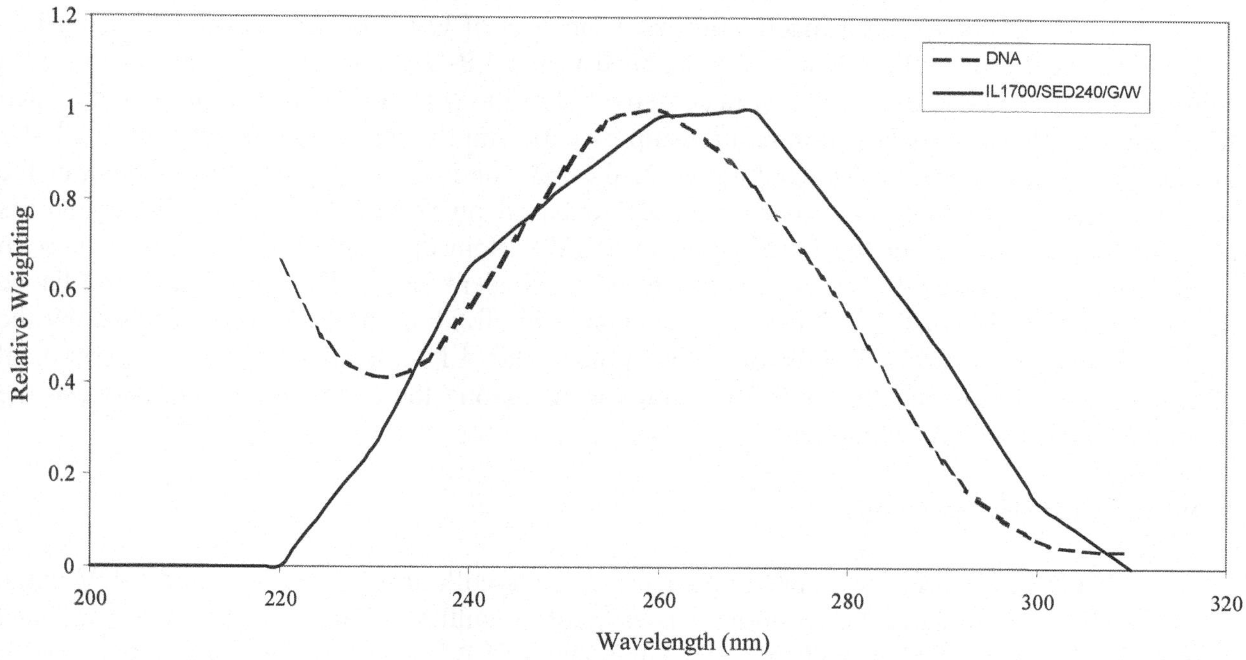

Source: Data from Jagger 1967.

Figure 3.4 Comparative germicidal weighting of "G" filter on the IL 1700 radiometer, SED 240 detector combination plotted with DNA absorbance spectrum (DNA spectra)

Pulsed-Ultraviolet Irradiance Determination

The radiometer method calculations and actinometer method calculations for P-UV have large discrepancies. Surprisingly, without mathematical analysis data are close, while after mathematical analysis they differ substantially. These differences may be a result of the different light emission mechanisms of pulsed versus continuous wave radiation that could affect the performance of radiometer and/or the light absorption of the actinometer. It would be expected that as long as photons are hitting the actinometer, it should accurately count them, no matter how they enter, unless there are multiple-photon hits occurring or non-documented photoproducts formed which would invalidate use of the quantum yields developed for one-hit kinetic models and known photoproducts. Another possible area of experimental error could be in the emission spectra measurements.

COMPARISON OF BIODOSIMETER TO RADIOMETER AND ACTINOMETER IRRADIANCE MEASUREMENTS

Biodosimetry

Biodosimetry was utilized as a tool to measure UV dose provided by the polychromatic lamps. The irradiance was calculated by dividing the dose obtained by the time of exposure. *B. subtilis* aerobic spores, obtained from American Type Culture Collection (ATCC #6051) were used as the biodosimeter. The spores were suspended in Metropolitan's biofilter effluent water and a relationship between log inactivation as a function of UV dose was developed using LP-UV collimated beam system. The UV irradiance in the LP-UV system was measured using a calibrated radiometer. The germicidal dose delivered in the MP- or P-UV system was estimated from this LP-UV – log inactivation relationship by exposing the spores to the MP-or P-UV light, calculating the log inactivation resulting from the MP- or P-UV exposure, and estimating the dose that the spores must have been exposed to based on the LP-UV results. This analysis assumes that an equivalent log inactivation in a UV system different from the LP-UV system could have only resulted from an exposure of equivalent dose. The germicidal irradiance delivered by the MP- or P-UV lamps was calculated by dividing the LP equivalent dose by the exposure time for the log inactivation data point. This LP equivalent delivered germicidal irradiance served as a benchmark in the comparisons among the technologies and between the irradiance measurement techniques.

Exposure to Ultraviolet Light

UV exposures were conducted under bench scale collimated beam MP- and P-UV units. 10 ml uridine actinometer, KI actinometer, *B. subtilis* culture (~10^8CFU/ml) were irradiated independently in identical petri dishes under the MP- and P-UV lamps. Targeted doses were 5, 10, and 20 mJ/cm^2. The estimated exposure times to obtain log reductions that were within the range of the LP-UV results were determined using radiometer readings and trial and error. All exposures were done in triplicate. Values for direct, incident, average and germicidal irradiance were measured as indicated above for radiometry and actinometry techniques.

The log inactivation response of *B. subtilis* resulting from exposure to a known dose of LP-UV was determined and the linear regression equation calculated. The delivered LP-UV germicidal dose was measured with a radiometer and adjusted for the absorbance effects of the *B. subtilis*-spiked water sample. The resulting regression equation for the UV dose – log inactivation relationship was: Log inactivation = 0.1736 × germicidal dose. Standard error of the estimated slope is 0.006765. Based on this linear regression equation, the delivered germicidal dose from the polychromatic, UV lamps resulting in each log inactivation was back calculated. Note that when approaching 4-log reduction (approximately dose of 20 mJ/cm^2), the log inactivation curve exhibited tailing. Thus, all the UV exposure experiments were designed for doses up to 20 mJ/cm^2 in order to remain in the region of linear UV dose - log reduction relationship.

Results comparing radiometer, uridine actinometer, KI actinometer and biodosimetry estimated irradiance values for direct, incident, average, and germicidal irradiance are shown in Figures 3.5 and 3.6 for MP-UV and P-UV respectively. "Direct (no correction)" is the irradiance estimated by the traditional quantification method without any correction. "Incident" is the total UV irradiance within the germicidal range (200-300nm) delivered to the water surface. "Average" is the representation of the total UV irradiance delivered to the completely mixed volume of water sample. "Germ" is the delivered germicidal irradiance. Since the direct biodosimetry estimation is a representation of the germicidal effective portion, the incident and average irradiance were not estimated from this method.

As illustrated in Figures 3.5 and 3.6, direct irradiance predictions from different methods are not consistent. For example, for the MP-UV work, there is about 40 percent difference between the highest estimation from KI and the lowest value predicted by uridine. After the

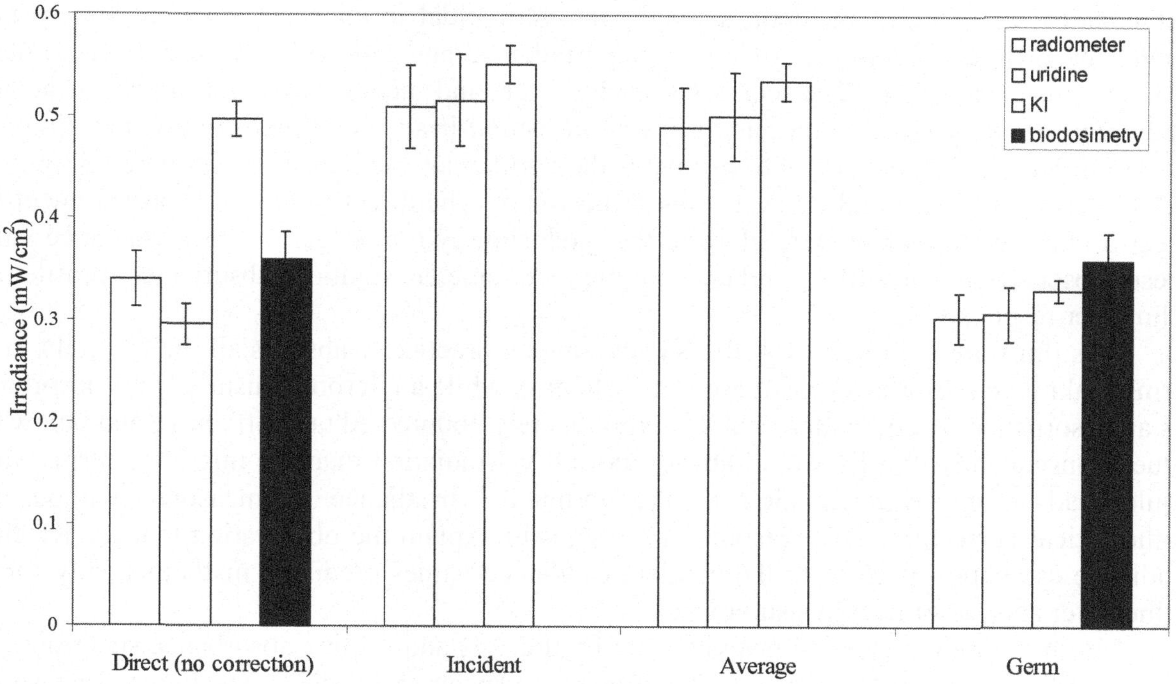

Figure 3.5 Comparison of ultraviolet irradiance measurements using radiometry, uridine and KI actinometry, and biodosimetry for the medium-pressure, ultraviolet lamp (95 percent confidence intervals shown)

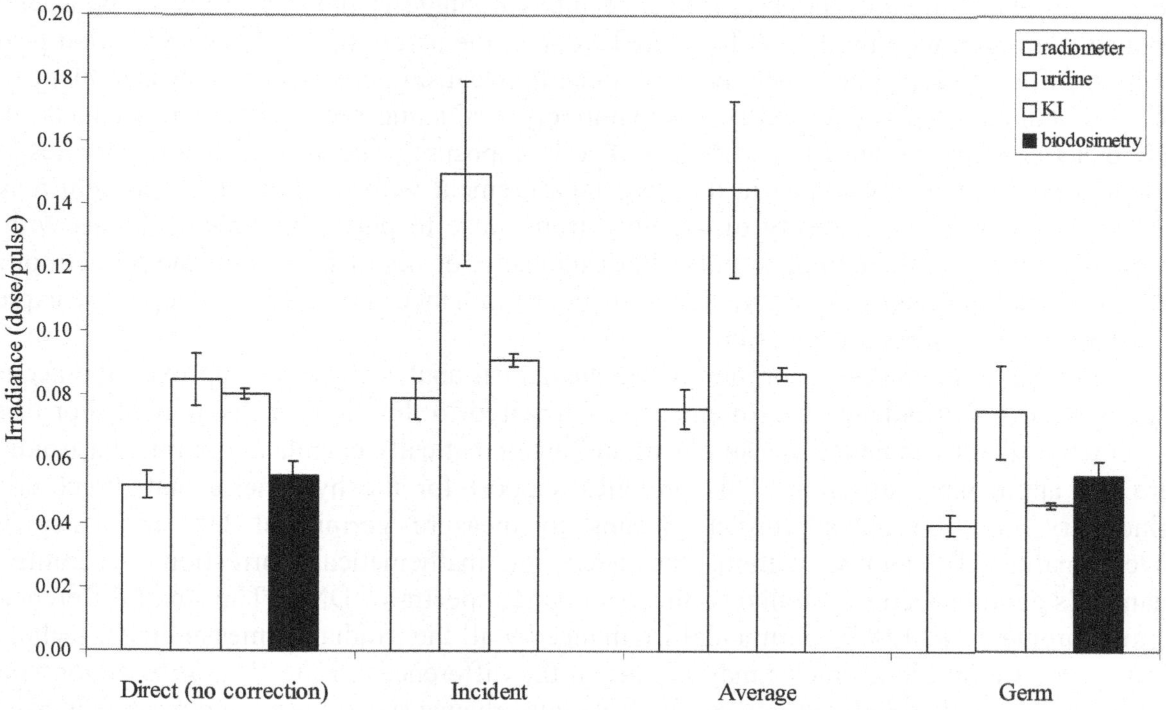

Figure 3.6 Comparison of ultraviolet irradiance measurements using radiometry, uridine and KI actinometry, and biodosimetry for the pulsed-ultraviolet lamp (95 percent confidence intervals shown)

mathematical corrections, incident, average, and germicidal irradiance values are less variable among the different techniques of measuring irradiance and fall within 95 percent confidence intervals of each other. The corrected radiometer and actinometer germicidal irradiance predictions are very close to the biodosimetry measured irradiance. Interestingly, the direct, no correction KI irradiance value over estimates the germicidal irradiance prediction by 33 percent and 42 percent for MP- and P-UV results respectively. The direct radiometer measurement and direct uridine measurement were close to the biodosimetry direct (germicidal) irradiance value. These observations are likely related to the actinometer solution absorbance profile and radiometer responsivity.

As illustrated in Figure 1.4, the KI actinometer practically absorbs all the UV light in the germicidal range (absorbance decreases after 280nm), while a microorganism's DNA absorbance has an absorbance spectra with a peak at approximately 260nm and tails off above and below this value. Consequently, the KI would absorb more UV irradiation than the microorganism, which would lead to an overestimation of the germicidal irradiance as measured without any mathematical correction. This reasoning also helps to explain the observation that the KI direct irradiance estimation is close to the incident irradiance values predicted mathematically for the actinometer and radiometer measurements.

Both the radiometer responsivity (See Figure 3.4) and uridine absorbance spectrum (See Figure 1.4) bare some resemblance to the absorbance spectrum of DNA (See Figure 1.2), so that even without mathematical correction, the measurement of irradiance is expected to be a reasonable representation of the germicidal irradiance. Note that one exception was the uridine actinometer data resulting from P-UV irradiation. The direct measurement of irradiance was greater for uridine than the other dose measurement methods, which differed from the comparatively lower value typically obtained using the uridine actinometer as illustrated in the MP-UV results. A notably larger confidence interval suggests the possibility of unusual data variance. Variation with the P-UV data may be due to the nature of the P-UV irradiation process resulting in unexpected photochemical reactions, thus causing errors in the measurement if the P-UV irradiance. This was tested by comparatively evaluating the degradation products of the uridine actinometry under LP-, MP-, and P-UV exposure. The results from these tests are presented below. Other variance in the dose measurement with the pulsed UV system may be due to variations in the intensity of the light from pulse to pulse, possible problems with the radiometer capturing the complete pulse flash intensity, or interactions with the KI actinometer that would lead to under or over prediction of light irradiance. These issues need to be explored in future work with pulsed UV systems.

For KI actinometry, incorporating mathematical analysis greatly improved the accuracy of germicidal UV irradiance measurement as compared to the direct measurement. For uridine actinometry, direct irradiance measurement and mathematically calculated germicidal irradiance were not appreciably different. This provides support for the hypothesis that direct uridine actinometry may be an appropriate means to measure germicidal UV irradiance from polychromatic, UV lamps, without the need for mathematical correction, assuming the organism's action spectra is similar to the absorbance spectra of DNA. The small differences in the measurement of MP-UV germicidal irradiance by all the irradiance measurement techniques as compared to the biodosimeter may be due to the differences in the *B. subtilis* action spectra (relative wavelength effectiveness) and the DNA absorbance spectra used in the calculation.

When examining the radiometer data, dose prediction from math analysis matches the uridine prediction, but also underestimates the germicidal dose for *B. subtilis*. However, the direct radiometer measurement predicts a similar value to that from biodosimetry. This suggests that the *B. subtilis* action spectra may have more similarity to the radiometer wavelength responsivity, which is modeled after the action spectra for *E. coli*. As illustrated in Figure 3.4, the radiometer responsivity curve spreads out wider than the DNA absorbance used, possibly resulting in the underestimation of predicted irradiance. In order to increase the accuracy of germicidal irradiance estimation, it would be preferable to use a germicidal weighting factor specific to the action spectra of the organism being investigated. However, for the purpose of cross comparison of different UV lamps, typically used to disinfect a multitude of microorganisms all with potentially different action spectra, a standard germicidal weighting factor such as the DNA absorbance spectra for predicting germicidal UV irradiance would be the best practical alternative available. For the purpose of UV lamp comparison, as long as the same germicidal weighting factor (e.g. DNA absorbance spectrum) is used, the estimation of germicidal irradiance employing mathematical analysis should always provide a value by which the lamps can be equitably compared.

URIDINE PHOTOHYDRATE BYPRODUCT ANALYSIS

The objectives of the uridine photohydrate byproduct analyses were to verify the quantitation of uridine and the corresponding photohydrate by UV absorbance and to search for other possible irradiation products of uridine that would cause concern for use of uridine to measure dose from polychromatic, UV lamps. The first phase of tests involved exposing a set of uridine samples to UV light from the MP-UV lamp at doses up to 30,000 mJ/cm^2. Analysis of these samples quantified the change in uridine absorbance across the UV spectrum after exposure to UV light and also determined where shifts in mass spectra may take place. A second phase of tests involved a detailed investigation into the possible production of unexpected photo-byproducts of uridine after exposure to both long-term doses of low- to moderate-UV light levels (LP- and MP-UV lamps) and doses of short-duration, high-intensity UV light (P-UV lamp).

Phase 1: Uridine Response to UV Light

Ultraviolet Absorbance Measurement

Uridine samples of 0.2 mM strength in Milli-Q® water were exposed to UV light emitted by the MP-UV lamp. Samples were given doses of 0 (control), 30, 300, 3,000, and 30,000 mJ/cm^2. Figure 3.7 illustrates the scanned absorbance spectra changes across the 200 to 400 nm range for the uridine solution after exposure to various UV light doses. The maximum absorbance wavelength and the corresponding absorbances are tabulated in Table 3.2. The decrease of absorbance between 260 to 262 nm is evident when the dose exceeds 300 mJ/cm^2, which suggests the photodecomposition of uridine. The absorption peak between 260 to 262 nm almost disappears with the highest dose (30,000 mJ/cm^2), indicating what may be the complete transformation of uridine to the proposed 5', 6' photohydrate.

To assure that mass spectra shifts would be detected, samples of uridine were given high doses of UV light (300 to 30,000 mJ/cm^2). Again, tests were conducted with the MP-UV lamp. Starting uridine concentrations ranged from 0.05 to 0.20 mM. Samples were analyzed by MS and the spectra obtained were background subtracted.

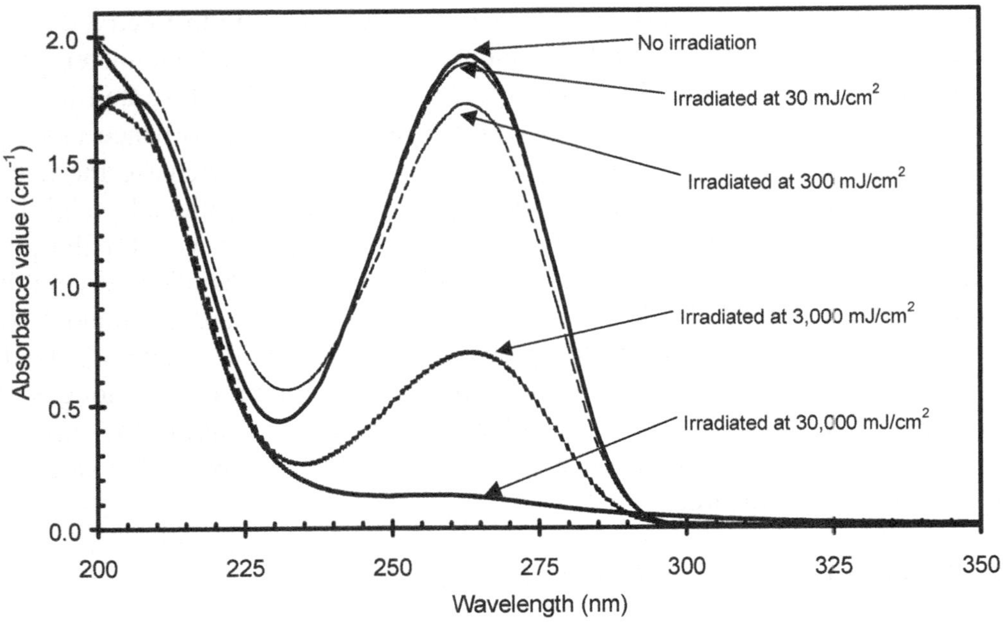

Figure 3.7 Absorbance of a 0.2 mM aqueous uridine solution exposed to increasing doses of ultraviolet light from a medium-pressure, ultraviolet lamp

Table 3.2
Ultraviolet absorbance of 0.2 mM aqueous uridine solution after exposure to increasing amounts of UV dose

UV Dose: (mJ/cm^2)	0	30	300	3,000	30,000
λ_{max} (nm)	260.2	261.6	261.0	261.8	260.6
Abs_{max}	2.296	2.281	2.138	1.162	0.075

Figure 3.8 presents spectra of the UV-treated 0.20 mM uridine. At 30,000 mJ/cm^2, the dominant peak for the 0.2 mM solution was the photohydrate. Only a minimal amount of uridine was left in solution (m/z 267). For the 0.05 mM solution (Figure 3.9) of uridine treated with UV, both uridine and the photohydrate were observed after the 1,000 mJ/cm^2 UV dose.

Mass Spectra Analysis

The output spectra in Figures 3.8 and 3.9 show and uridine monomers (m/z 267), dimers (m/z 511) and trimers (m/z 755) and photohydrate monomers (m/z 285), dimers (m/z 528) and trimers (m/z 791). In Figure 3.9, two additional peaks (m/z 529 and m/z 773) are noted, possibly correspond to [2MUridine+MH2O+Na]+ and [3MUridine+MH2O+Na]+, respectively. However, the possibility of these two peaks arising from products other than the photohydrate is not completely ruled out and high resolution experiments would be necessary to confirm their identities. Figure 3.8 reveals that the 0.20 mM uridine solution exposed to 30,000 mJ/cm2

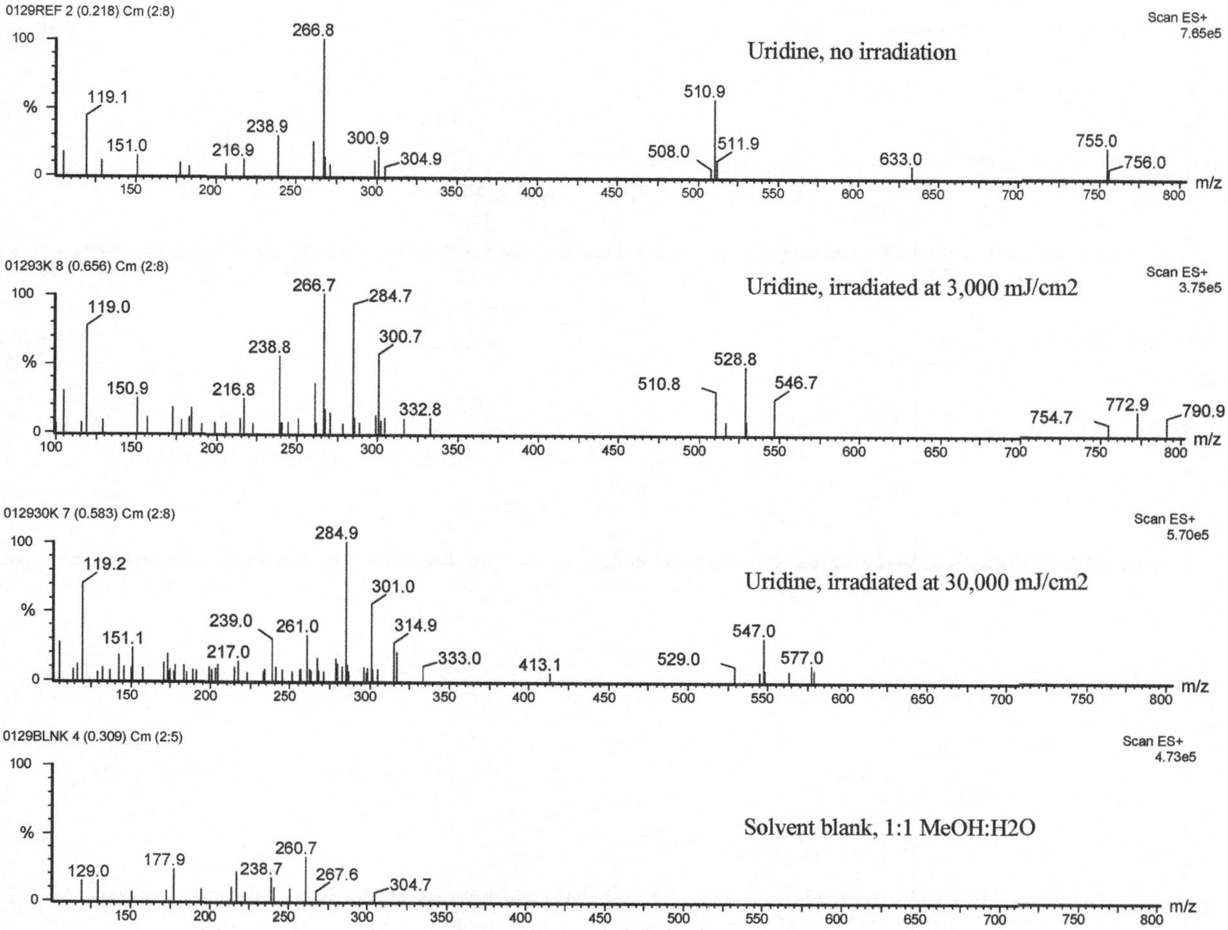

Figure 3.8 Mass spectra of 0.20 mM uridine solution with and without exposure of up to 30,000 mJ/cm^2 UV dose and the methanol solvent blank

creates two new peaks when compared to the blank uridine spectra. These peaks (m/z 315 and m/z 577) are unknown and are not present in the sample spectra.

Phase 2: Detailed Photobyproduct Investigation

This testing was conducted to (1) compare the full scan mass spectra of uridine exposed to light from the different UV lamps; (2) compare results when using different methods of introducing samples to the MS; and (3) to quantitate uridine and the photohydrate by mass spectrometry. A Micromass AutoSpec-oaTOF magnetic sector instrument was used for the analysis.

After extensive methods development, six, 0.012 mM uridine samples in Milli-Q® water were exposed to approximately 600 (sample 1) and 1,600 (sample 2) mJ/cm^2 of UV light from the LP-, MP-, and P-UV lamps. Changes observed in uridine absorbance are summarized in Table 3.3.

Irradiated samples were scanned from 200 nm to 400 nm with a scanning spectrophotometer. The following equation shows how the extinction coefficient of uridine was divided into the absorbance to quantify uridine concentration in the samples:

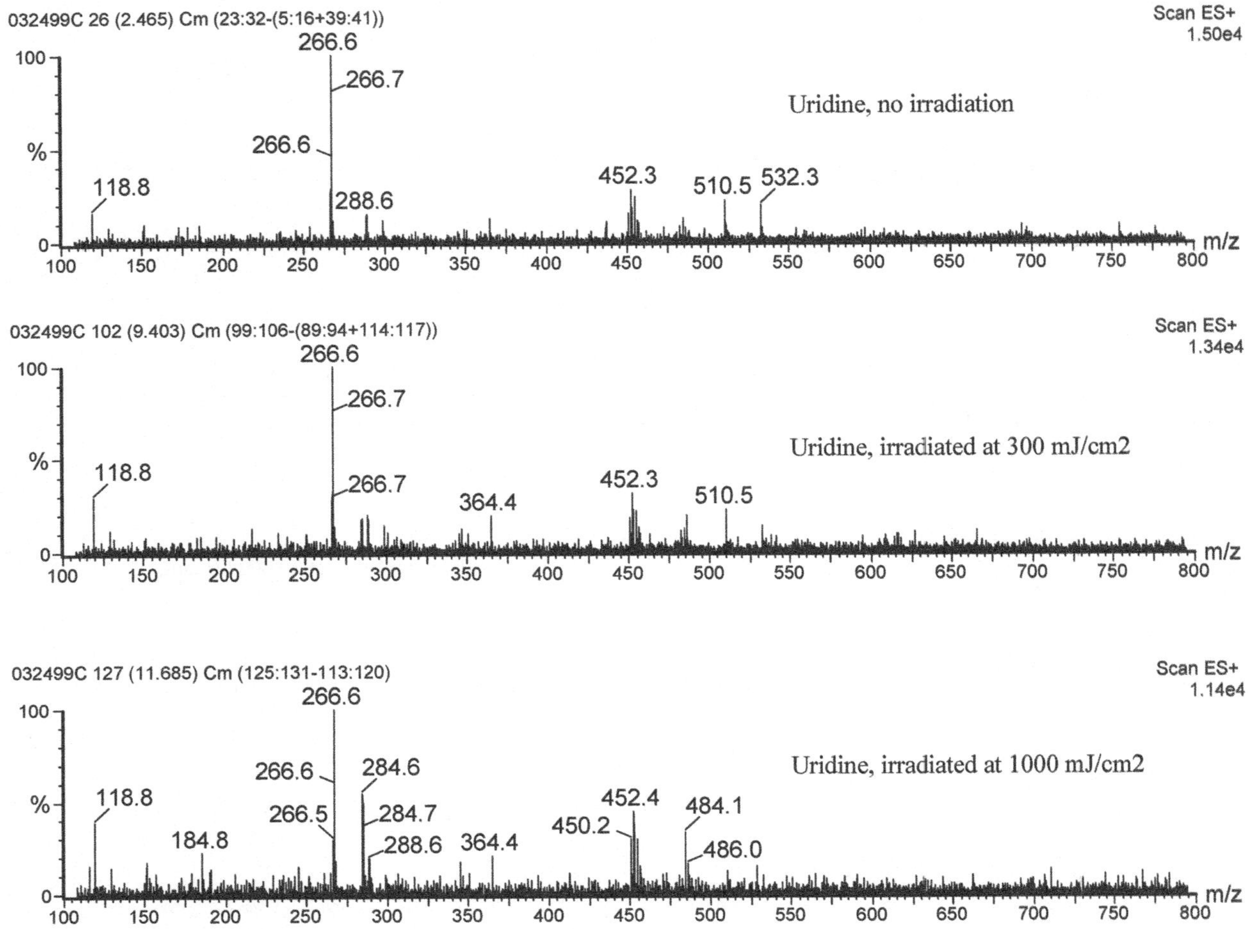

Figure 3.9 Mass spectra of 0.05 mM uridine solution with and without exposure of up to 1,000 mJ/cm^2 UV dose

$$\text{Concentration of uridine} = \frac{\text{absorbance at 262nm}}{9680\ \text{L/mol} * \text{cm (1cm)}} \tag{3.1}$$

Table 3.3 presents the calculated values of the uridine concentration. The 0.012 mM uridine blank solution was prepared in two separate batches for this testing. One blank was used for the LP-UV samples and one was used for the MP- and P-UV samples. There was notable difference in UV absorbance when the same reference solution was measured either at different times of the day, or on different days. This is shown by the blank absorbance for MP- and P-UV tests changing from 0.117 to 0.132 on the same day and 0.117 to 0.149 on different days. This variability may have been due to instrument error or actinometer instablity. Thus, the project team conducted a detailed evaluation of actinometer stability which is summarized in the following section.

Table 3.3
Irradiated uridine samples

UV Technology	Uridine Absorbance* (Blank)	Uridine Absorbance* (Sample)	Reduction of Uridine Absorbance* (Percent)	Reduction of Uridine Absorbance* ($\log_{10}$)	Uridine Concentration Blank (mM)	Uridine Concentration Sample (mM)
LP-UV Sample 1†	0.135	0.061	54.7	0.34	0.0139	0.0063
LP-UV Sample 2	0.135	0.041	69.9	0.52	0.0139	0.0042
MP-UV Sample 1†	0.117	0.062	47.1	0.28	0.0120	0.0064
MP-UV Sample 2	0.132	0.037	72.3	0.56	0.0136	0.0038
P-UV Sample 1†	0.149	0.077	48.3	0.29	0.0154	0.0079
P-UV Sample 2	0.137	0.043	68.5	0.50	0.0142	0.0044

* Absorbance measured at 262 nm

† Two different uridine solutions were prepared (LP-UV challenges used one blank solution and the MP-UV and P-UV challenges used a second blank solution), and each lamp was challenged with the uridine on different days

Full-Scan Analysis

Due to low initial uridine concentration, full scan samples were injected into the MS directly without addition of methanol. The mass scan range was 100-800 m/z and the mass spectra obtained were background subtracted (see Figure 3.10). The spectrum of uridine solution A gave an unusual dominant peak at m/z 116, which is not seen from uridine solution B and its origin is unknown. When displayed from m/z 120-800 (see Figure 3.11), a better view of the uridine spectra for the blank solution and irradiated samples can be seen. Figures 3.12 and 3.13 show the uridine spectra from the MP- and P-UV experiments, respectively.

Other than the two peaks from uridine and the photohydrate (267 and 285 m/z, respectively) most of the peaks on Figures 3.10 through 3.12 are low in intensity and narrow in peak width. This indicates that they are more likely to be random ions than real ion peaks. There were no obvious high mass ions in any of the samples that suggest the covalent dimer and polymer of uridine. These species were either not formed or beyond detection limit of the mass spectrometer due to the low concentration of starting uridine solution.

Quantitative Analysis

All samples were diluted in equal volume of methanol. Two injection methods were used: (1) caffeine standard combined with the samples prior to injection; and (2) caffeine standard and samples introduced to the ESI in alternate injections (i.e., caffeine, then first sample, then caffeine, then second sample, etc.). The combined method was used to evaluate if instrument variability may play a significant role in uridine and photohydrate detection. The alternating injection method was used to understand if possible cross-dimers or cross-trimers between the sample and the caffeine standard may be avoided. Also, with the caffeine standard added to the samples, better compensation may be made to overcome instrument variability.

(A) Spectra for blank solution A

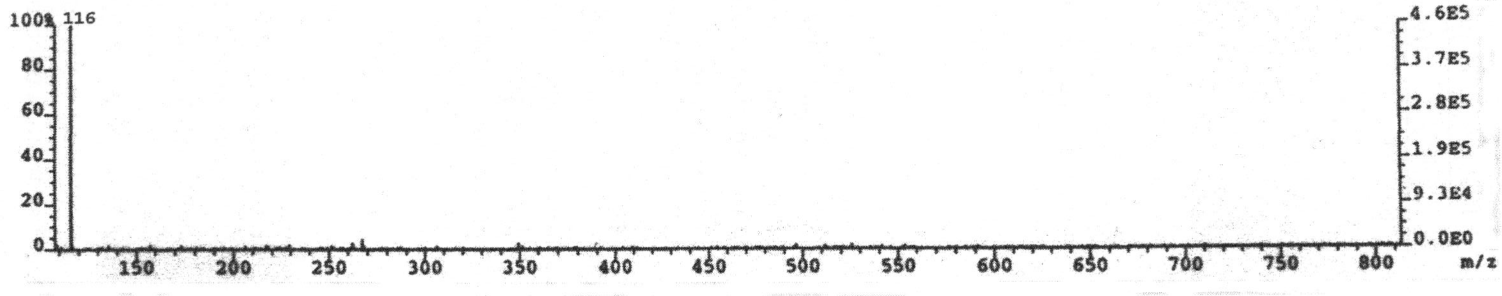

(B) Spectra for low-pressure, ultraviolet sample 1

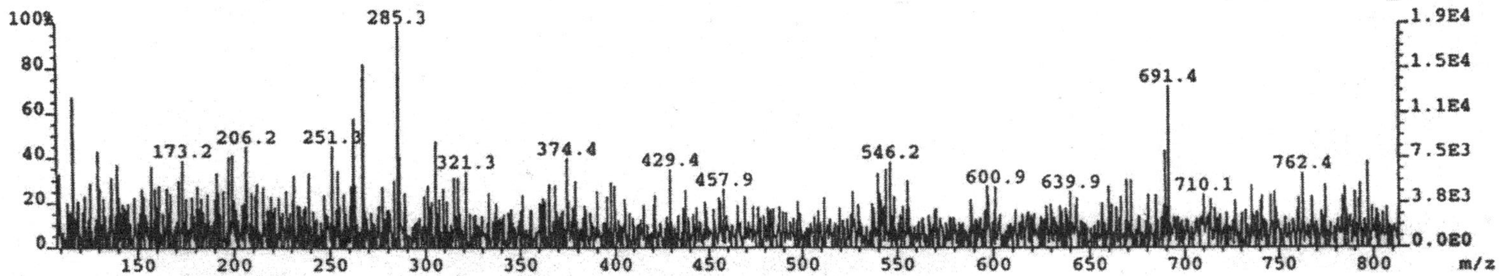

(C) Spectra for low-pressure, ultraviolet sample 2

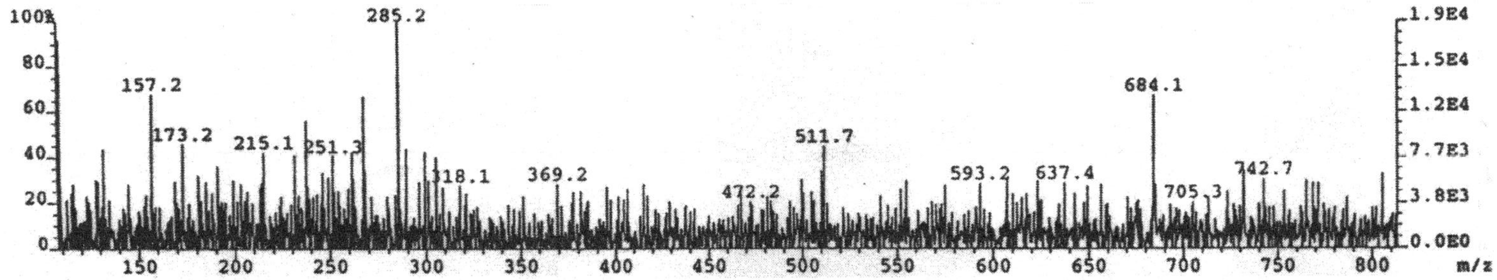

Figure 3.10 Mass spectra of uridine blank solution A and low-pressure, ultraviolet samples (mass range from 100 to 800 m/z)

(A) Spectra for blank solution A

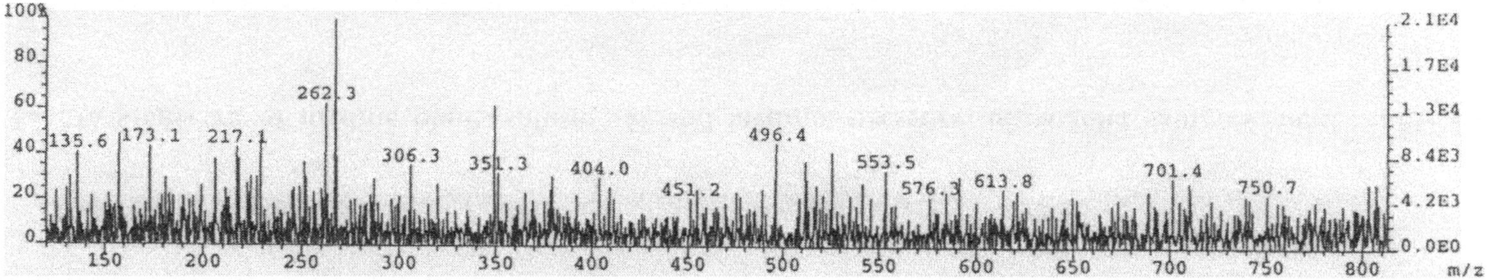

(B) Spectra for low-pressure, ultraviolet sample 1

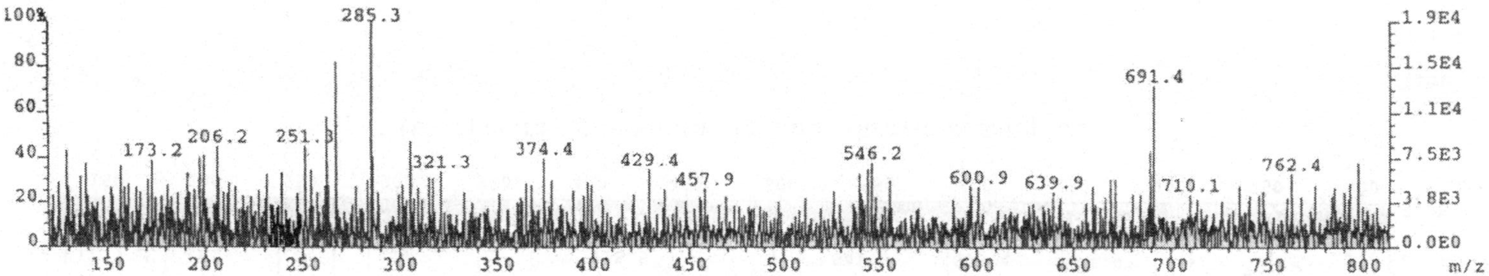

(C) Spectra for low-pressure, ultraviolet sample 2

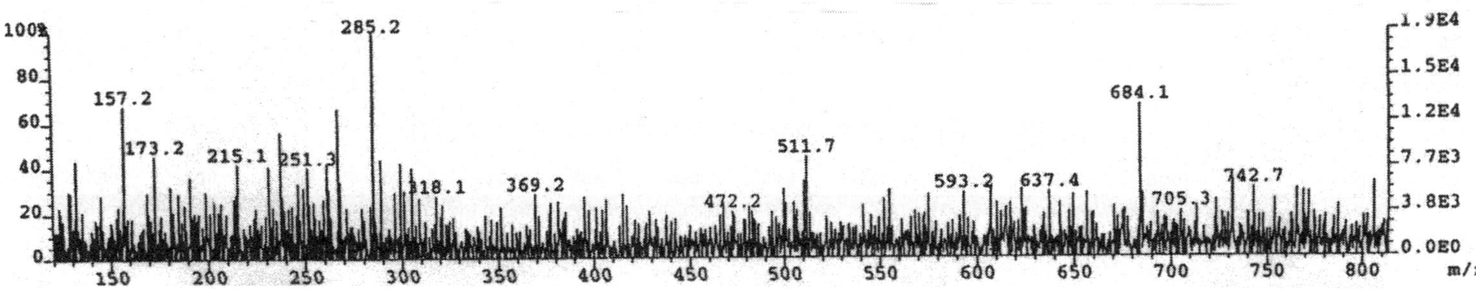

Figure 3.11 Mass spectra of uridine blank solution A and low-pressure, ultraviolet samples (mass range from 120 to 800 m/z)

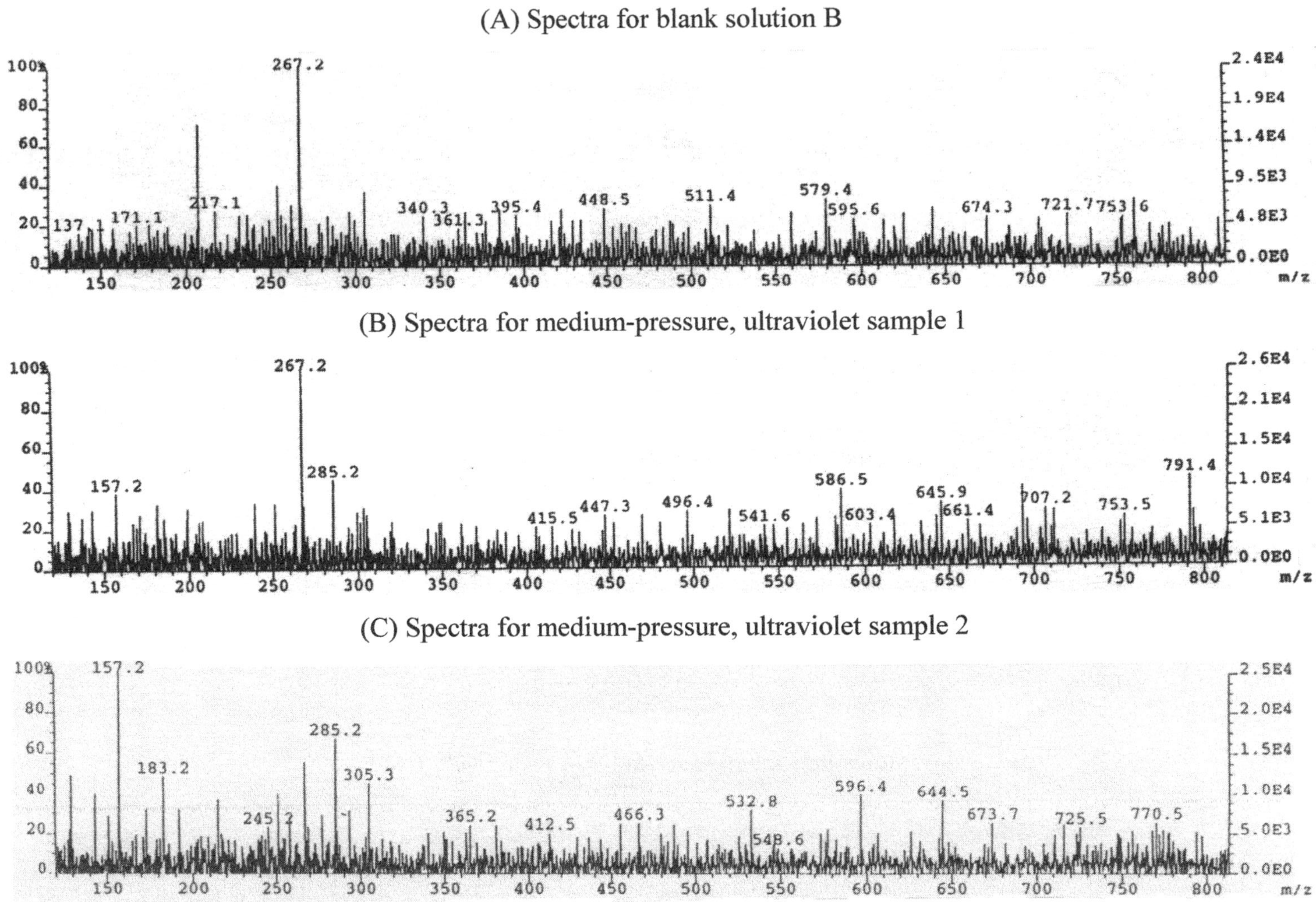

Figure 3.12 Mass spectra of uridine blank solution B and medium-pressure, ultraviolet samples (mass range from 100 to 800 m/z)

(A) Spectra for blank solution B

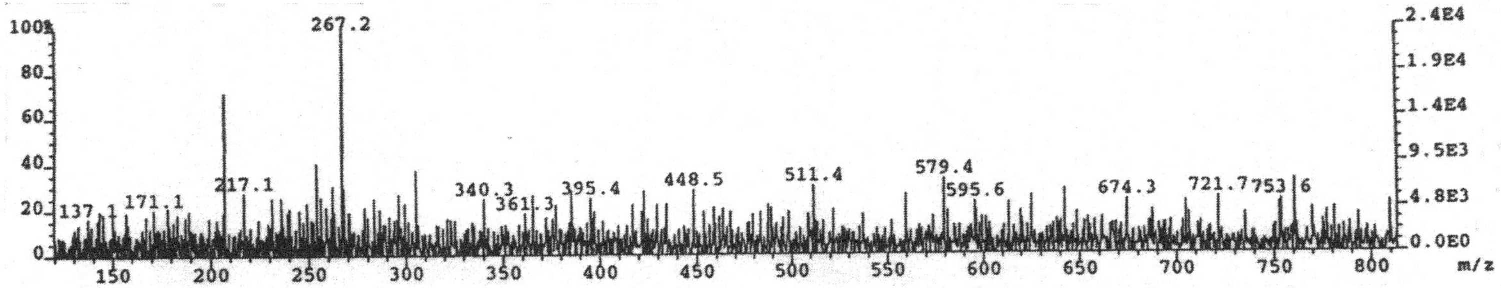

(B) Spectra for pulsed-ultraviolet sample 1

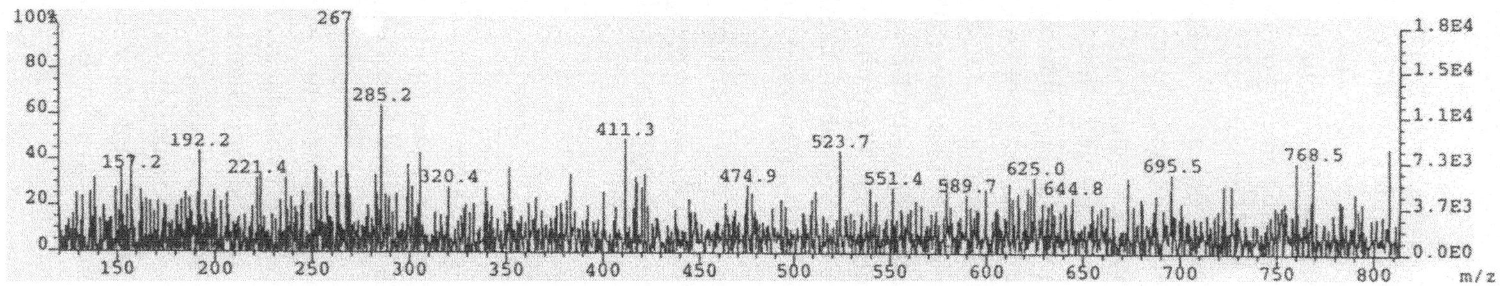

(C) Spectra for pulsed-ultraviolet sample 2

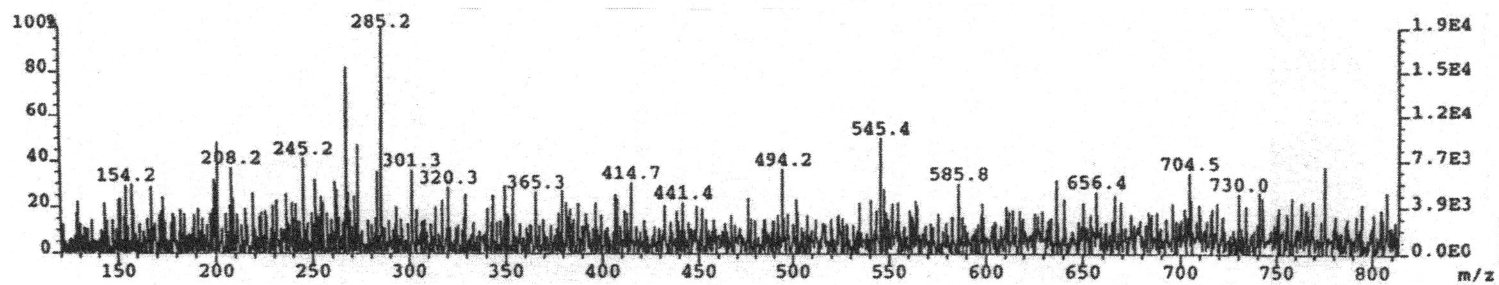

Figure 3.13 Mass spectra of uridine blank solution B and pulsed-ultraviolet samples (mass range from 100 to 800 m/z)

Quantitative analysis of uridine and the photohydrate of the six samples was carried out using the selected ion recording (SIR) mode by monitoring the masses corresponding to the sodium adducts of uridine ($[M+Na]^+$, 267 m/z), the photohydrate ($[M+Na]^+$, 285 m/z), and the caffeine standard ($[M+MeOH+Na]^+$, 249 m/z). The MS analyses were carried out both with the combined and alternating injection methods.

The goal for these tests was to produce the photohydrate after long exposure to UV light. Similar to the samples described above, target doses were 600 (sample number 1) and 1,600 (sample number 2) mJ/cm^2 for all three lamps. Results from combining the sample and standard injections into the MS are presented in Table 3.4. P-UV sample 2 showed a low mass balance of 69 percent, possibly indicating that there are other products formed in addition to the photohydrate. The mass balance for the LP-UV samples 1 and 2 showed 98 percent and 113 percent recoveries, respectively; the mass balance for the MP samples 1 and 2 showed 96 percent and 89 percent recoveries, respectively; and, the mass balance for the P-UV samples 1 and 2 showed 83 and 69 percent recoveries, respectively. These results may indicate that unexpected uridine photobyproducts have been formed when either (1) a long-duration, low- to moderate-intensity UV light is applied and/or (2) high-intensity, short-duration pulses of UV light are applied. It is possible that these products may not be detected by MS because (1) they are formed at a low concentration which is beyond instrument detection and/or (2) they do not ionize under the electrospray positive ion mode. One possible byproduct which fits into this criteria is carbon dioxide which is not detectable by electrospray.

Table 3.5 summarizes MS test results when alternating injections were made. Compared to Table 3.4, it is evident that mass balances decreased substantially. This may indicate that the alternating injection method may not be the most appropriate method for monitoring the uridine photohydrate.

After interpretation, MS analyses of the irradiated samples show no significant differences when comparisons are made between the samples exposed to UV light from the three lamps. The results suggest that the expected photohydrate was the only significant photohydrate byproduct formed.

Summary

Table 3.6 summarizes the uridine concentrations calculated from the UV absorbance measurements and the MS analyses using the two injection methods. In general, UV absorbance measurements were more similar with MS/combined injection results than the MS/alternating injection results.

After inspecting all results from mass balances of uridine after exposure to UV light, there is a possibility that unknown photohydrate byproducts are formed. This may be the case when the UV dose is very high from a low- to moderate-intensity source and/or when it is from a short-duration, high-intensity, P-UV lamp source. These products may not be detected by MS because due to either their low concentration or because they can not be detected by the methods used in this study. However, this result should not be a concern when UV doses in the disinfection-level range (i.e., less than 300 mJ/cm^2) need to be quantified.

Table 3.4
Uridine and photohydrate concentrations measured after combining sample injection with the standard solution (starting uridine concentration = 0.012 mM)

Sample Identification	Uridine [U] (mM)	Photohydrate [H] (mM)	[U] + [H] (mM)	Mass Balance (Percent)
Blank for LP-UV	0.0108	ND	0.0108	90
LP-UV, Sample 1	0.0059	0.0058	0.0117	98
LP-UV, Sample 2	0.0049	0.0087	0.0136	113
Blank for MP-UV and P-UV	0.0120	ND	0.0120	100
MP-UV, Sample 1	0.0068	0.0047	0.0115	96
MP-UV, Sample 2	0.0043	0.0064	0.0107	89
P-UV, Sample 1	0.0071	0.0029	0.0100	83
P-UV, Sample 2	0.0044	0.0039	0.0083	69

ND = not detected
LP-UV = low-pressure UV test
MP-UV = medium-pressure UV test
PU-V = pulsed-UV test

Table 3.5
Uridine and photohydrate concentrations measured after alternating sample injection with the standard solution (starting uridine concentration = 0.012 mM)

Sample Identification	Uridine [U] (mM)	Photohydrate [H] (mM)	[U] + [H] (mM)	Mass Balance (Percent)
Blank for LP-UV	0.0078	ND	0.0078	65
LP-UV, Sample 1	0.0043	0.0054	0.0097	81
LP-UV, Sample 2	0.0035	0.0089	0.0124	103
Blank for MP-UV and P-UV	0.0099	ND	0.0099	83
MP-UV, Sample 1	0.0053	0.0040	0.0093	78
MP-UV, Sample 2	0.0033	0.0070	0.0103	86
P-UV, Sample 1	0.0065	0.0027	0.0092	77
P-UV, Sample 2	0.0035	0.0042	0.0077	64

ND = not detected
LP-UV = low-pressure UV test
MP-UV = medium-pressure UV test
P-UV = pulsed-UV test

STABILITY OF ACTINOMETER SOLUTIONS UNDER OPERATING CONDITIONS

Dose measurement using actinometry is predicated on the stability of the photochemical reactions for a long enough period to quantify the photon based changes in the parent actinometer compound. Previous studies using uridine have determined that the compound undergoes a 5', 6' photohydration reaction upon UV radiation (Sinsheimer and Hastings 1949,

Table 3.6
Percent difference in measurement of uridine concentration when comparing the ultraviolet-absorbance method to the mass-spectrometer injection methods

		Combined Injection		Alternating Injection	
Sample Identification	Concentration Measured by UV Absorbance (mM)	Conc. by MS Scan (mM)	Difference from UV Absorbance* (diff.)	Conc. by MS Scan (mM)	Difference from UV Absorbance* (diff.)
Blank (for LP-UV tests)	0.0139	0.0108	22.3	0.0078	43.9
LP-UV Sample 1	0.0063	0.0059	6.3	0.0043	31.7
LP-UV Sample 2	0.0042	0.0049	16.7	0.0035	16.7
Blank (for MP-UV and P-UV tests)	0.0120†	0.0120	ND	0.0099	17.5
MP-UV Sample 1	0.0064	0.0068	6.3	0.0053	17.2
MP-UV Sample 2	0.0038	0.0043	13.2	0.0033	13.2
P-UV Sample 1	0.0079	0.0071	10.1	0.0065	17.7
P-UV Sample 2	0.0044	0.0044	ND	0.0035	20.5
Average:			9.4		22
Standard deviation:			7.8		10

ND = No difference
* Percent differences shown are absolute value
† Four measurements were made (see Table 3.2) but only one was compared to MS

Moore and Thompson 1955, von Sonntag and Schuchmann 1992, Wang 1962) as shown in Figure 3.14. The photohydration product does not absorb at the peak absorption wavelength of uridine, which is 262nm (Wang 1962). This reaction follows first-order kinetics (Moore and Thompson 1955, Wang 1962). The uridine photoproduct of UV irradiation was found to be unstable under certain conditions and may revert back to the original uridine or to transform to a new substance (Sinsheimer 1954). These conditions include acidification or heating to reverse the photoproduct formation or alkaline conditions to pH 13, converting the photoproduct to substances other than uridine, (Sinsheimer and Hastings 1949, Sinsheimer 1954). Both irradiated and unirradiated KI actinometer was found to be relatively unstable under ambient conditions (Rahn 1997, Rahn 1993). Iodide solutions can undergo spontaneous formation of triiodide due to thermal oxidation which may lead to overestimation of dose (Rahn 1997). Triiodide was found to undergo a gradual decrease with time. Thus an investigation into the stability of uridine and KI actinometers as used in the determination of UV dose in disinfection studies was initiated to recommend proper storage and handling of the actinometers for accurate and unbiased light measurement applications.

To evaluate the stability of the actinometers solutions of uridine and KI were irradiated under different doses of UV light from LP-UV lamps. Solutions were monitored over time under different environmental conditions (light/dark, cold/ambient/warm) and the impact of the storage time and conditions on the dose measurement were determined (See Appendix C for experimental details).

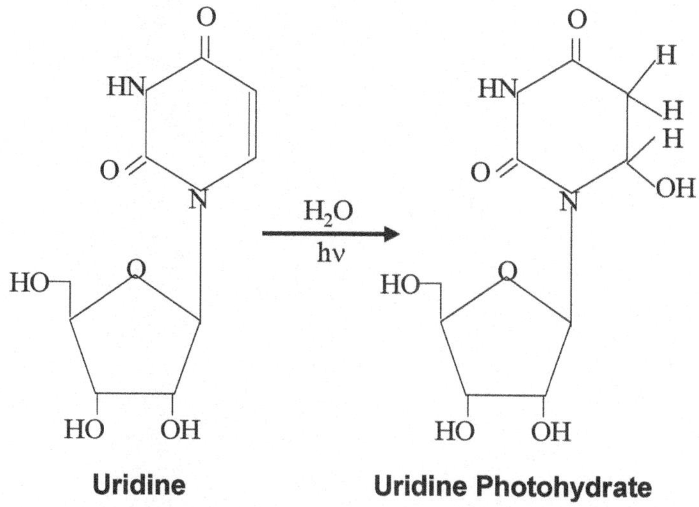

Figure 3.14 Chemical structure of uridine and the uridine photohydrate

It was found that uridine is a relatively stable actinometer. For most accurate use of uridine actinometer it is most desirable to keep any irradiated and unirradiated actinometer under cold storage. Unirradiated uridine actinometer can be left under room condition for at least 7 days but it is not recommended to leave any irradiated sample at room temperature. In addition, the higher the UV dose, the more reactive the uridine may be. Thus good laboratory practice would indicate that uridine should be analyzed as soon as possible following irradiation, and that if storage is necessary, cold storage should be utilized.

The iodide/iodate actinometer is relatively unstable and precautions must be taken in handling and storage. These precautions include: (1) making fresh solution on each day of use; (2) recording of all absorbance measurements in reference to a control (unexposed) measurement of the solution; (3) avoiding prolonged storage; (4) storing in cold, dark conditions; and (5) anticipation of significant operational error at lower UV irradiation doses.

CHAPTER 4
UV DOSE / LOG-INACTIVATION RELATIONSHIPS

BACILLUS SUBTILIS

Testing conducted with the aerobic spore-forming bacteria *B. subtilis* was carried out according to the matrix presented in Table 4.1. Initial tests were conducted in buffered laboratory water to develop a baseline dose-response relationship of the bacteria. These baseline tests were conducted with all three lamp types over a wide range of targeted UV doses (5 dose steps between 10 to 100 mJ/cm^2). Delivered UV dose measurement was calculated using direct radiometer readings (providing mW/cm^2 values of UV irradiance), solution absorbance values, depth of solution, and measurement of exposure time. The UV dose calculation procedure is detailed in Chapter 2. Raw data for all tests is presented in Appendix D.

Figure 4.1 presents the response of *B. subtilis* spores, while suspended in buffered laboratory water, to light from a LP-UV lamp. This figures shows data points collected during the study, the linear regression through those data, and the calculated 95 percent prediction limits around the data. All tests were conducted at UV doses less than 110 mJ/cm^2 and several dose-steps were evaluated so that an adequate UV-dose, microorganism response could be captured.

The data indicate that at the lower UV doses (10 and 20 mJ/cm^2), significantly more rapid disinfection kinetics occurs when compared to the higher UV doses (from 40 to 100 mJ/cm^2). Variability in the data was significantly different between these two ranges as well. Throughout the lower UV doses, the linear regression was characterized by an R^2 value of 0.95. This implies that 4 log inactivation of spores would be achieved at 23 mJ/cm^2. The upper UV dose range had a linear regression with an R^2 of 0.60 and indicates that only 1-log greater inactivation of spores would be achieved after a 4-fold increase in UV dose. This break in inactivation kinetics, indicating a "tailing" in the data, may have been due to the existence of a subpopulation of spores in the sample that were more resistant than others, or spores that exhibit ranges of difference in their responsivity to UV, possibly due to age. However, the research team did not investigate possible reasons behind why there would be this tailing, or 'break' in the dose response. For comparison between the different waters and UV treatments, the primary dose response curve, up to a 4 log inactivation of spores, was used to represent the relationship between UV dose and inactivation of the spores.

Table 4.1
***Bacillus subtilis* experiment matrix**

Parameter	Levels	Values
Water type	4	Buffered laboratory water California natural water matrix Burlington natural water matrix Wilmington natural water matrix
Lamp type	3	Low-pressure, medium-pressure and pulsed
Target UV doses*	5	5, 10, 15, 25, and 40† mJ/cm^2

* A radiometer estimated exposure time before conducting experiments, all UV doses were repeated three times for each lamp.

† Initial tests in buffered water were done at 10, 20, 40, 70, and 100 mJ/cm^2

After completion and review of buffered water tests, tests were conducted in the three natural surface waters. Tests were conducted with the LP-, MP- and P-UV lamps. Figures 4.2, 4.3, and 4.4 show data from the natural water testing. Chapter 2 details the water quality information collected during these tests. Test data were collected for these tests at slightly different dose ranges than the buffered water tests. The doses were lower in the natural water tests than for the buffered water tests. The lowering of the UV doses was done in an attempt to appropriately capture the region of more-rapid inactivation kinetics.

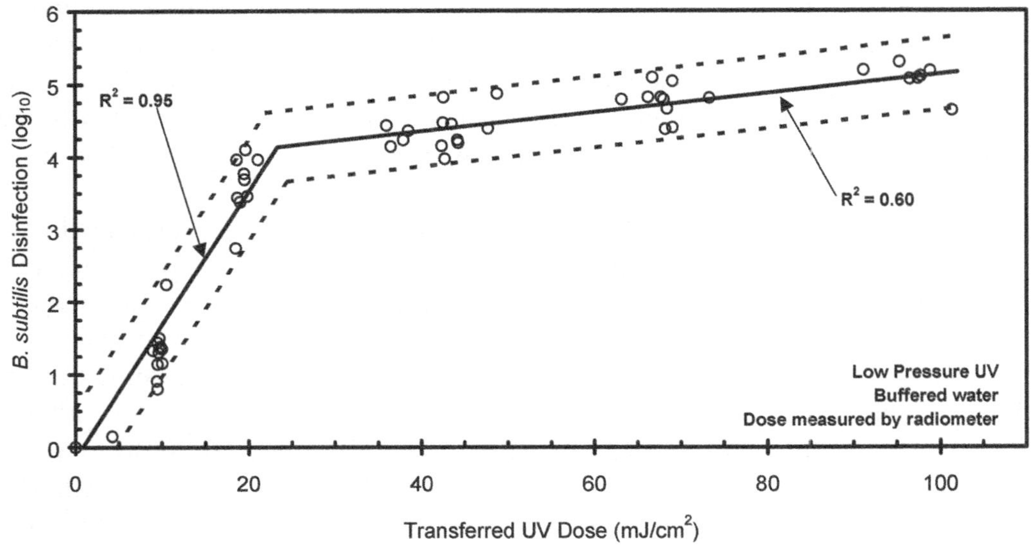

Figure 4.1 Low-pressure, ultraviolet inactivation of *Bacillus subtilis* in buffered water plotted with 95 percent confidence interval

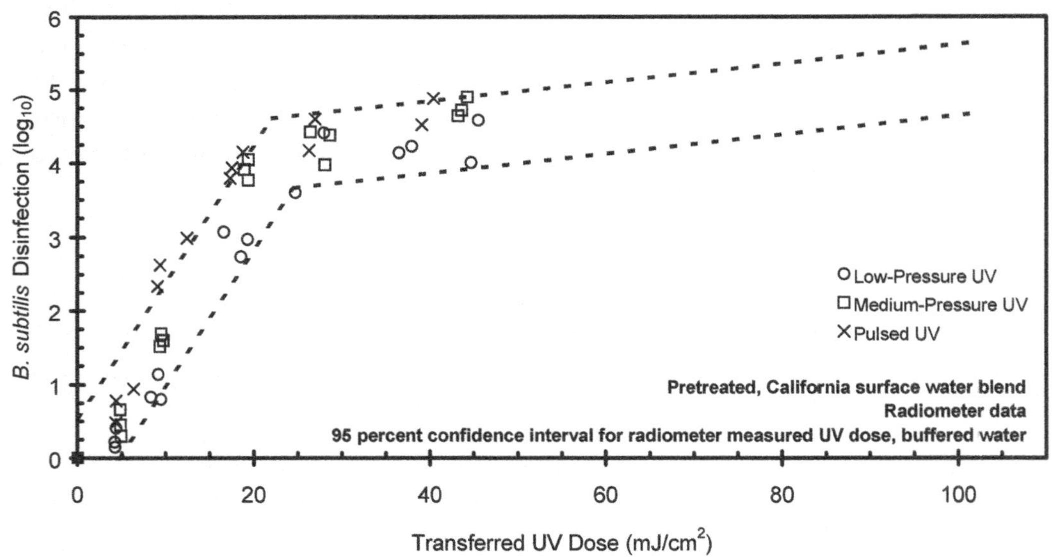

Figure 4.2 Disinfection efficiency of *Bacillus subtilis* by all three ultraviolet lamp types in pretreated, California surface water blend (plotted against the low-pressure, buffered water, 95 percent confidence interval)

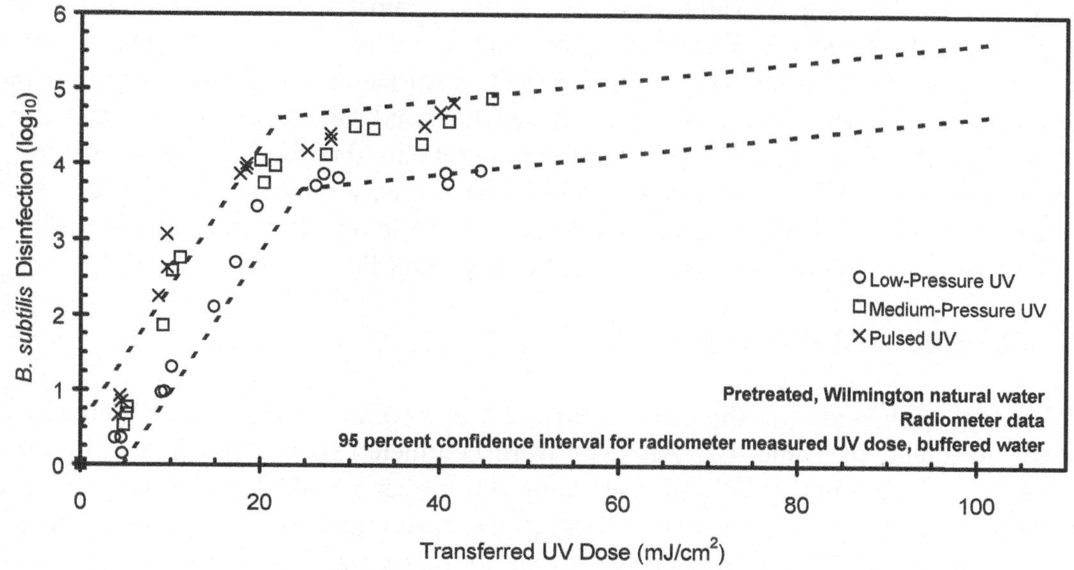

Figure 4.3 Disinfection efficiency of *Bacillus subtilis* by all three ultraviolet lamp types in water from the Sweeney Water Treatment Plant in Wilmington, NC (plotted against the low-pressure, buffered water, 95 percent confidence interval)

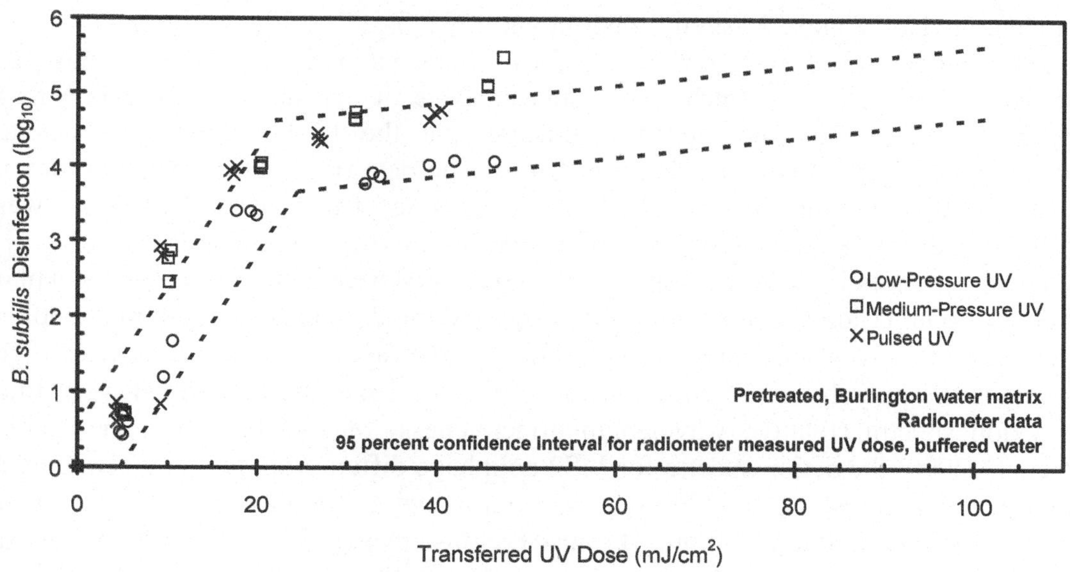

Figure 4.4 Disinfection efficiency of *Bacillus subtilis* by all three lamp types in surface water from the Ed Thomas Water Treatment Plant in Burlington, NC (plotted against the low-pressure, buffered water, 95 percent confidence interval)

Results of the natural water tests are plotted against the 95 percent prediction limits generated in the buffered-water tests described above. The prediction limits are offered to compare results in the natural waters (higher light absorbance and light scattering, increased organic loading) with the more controlled tests in buffered laboratory water. Data from these tests show that the dose-response for all three lamps in the three different natural waters appeared to fall inside the buffered-water prediction limits. Across the tests in all the waters, however, there were a few data points from each of the lamps that fell outside the prediction limits. The raw *B. subtilis* test data is presented in Appendix D.

MALE-SPECIFIC COLIPHAGE MS-2

Testing conducted with the virus surrogate MS-2 coliphage was carried out according to the matrix presented in Table 4.2. All tests were conducted in a blend of California surface waters. Tests were conducted with all three lamp types over a wide range of targeted UV doses (5 dose steps between 10 to 75 mJ/cm^2). UV dose measurement was performed similar to that described for the *B. subtilis* experiments. Raw data for these tests are offered in Appendix D.

Figure 4.5 presents the response of MS-2 coliphage, while suspended in the blend of California surface waters, to light from all three UV lamps. Linear regressions of the LP-, MP-, and P-UV lamp data are characterized by R^2 values of 0.96, 0.99, and 0.98, respectively. The slope of the regression line indicates that a LP-UV dose of 32 mJ/cm^2 or more would provide 2-log inactivation of MS-2. For the MP- and P-UV lamps, this same level of microbial control was demonstrated at 28 and 22 mJ/cm^2, respectively.

The trends in the dose-response data for all three lamps is somewhat similar. However, after characterizing data with linear regression analysis, it appears that there is a ranking in disinfection kinetics. This ranking may indicate that the LP-UV lamp provides a lower inactivation per UV dose than the other two lamps. The data from the P-UV lamp may indicate the most rapid disinfection kinetics. The linear regression of the data for all lamps was linear across all doses examined indicating good precision in experimental methods.

The differences in the dose-response among the different lamp types may be explained by the difference in emission spectra among the lamps and the dose measurement methodology. The doses for each lamp were measured using the bias of the radiometer "G" filter (See Figure 3.4) which weighs the polychromatic emissions according to a specific germicidal weighting based on the action spectra (relative wavelength effectiveness) of *E. coli*. However, according to Linden et al. (2000) the action spectra for MS2 coliphage differs from that of *E. coli* and appears to favor wavelengths toward the 270 nm range and below 230 nm as more efficacious than the wavelengths between 250 and 270 nm. Because the dose measurement methodology used would not properly account for these regions of wavelength effectiveness, the estimated delivered doses may have been underestimated. This difference in response is inherent in the dose measurement method used. Future work should investigate the use of different dose weighting schemes and analyze dose response using different wavelength germicidal biases. This exercise exemplifies the importance of detailing how dose is measured when using polychromatic, UV lamps. It also illustrates both the need for development of a universal standard for measuring UV dose from polychromatic lamps and the problems that adopting such a standard would pose when trying to evaluate a range of microbes with different action spectra.

Table 4.2
MS-2 coliphage experiment matrix

Parameter	Levels	Values
Water type	1	California natural water matrix
Lamp type	3	Low-pressure, medium-pressure and pulsed
Target UV doses*	5	10, 20, 30, 50, and 75 mJ/cm^2

* A radiometer estimated exposure time before conducting experiments, all UV doses were repeated three times for each lamp.

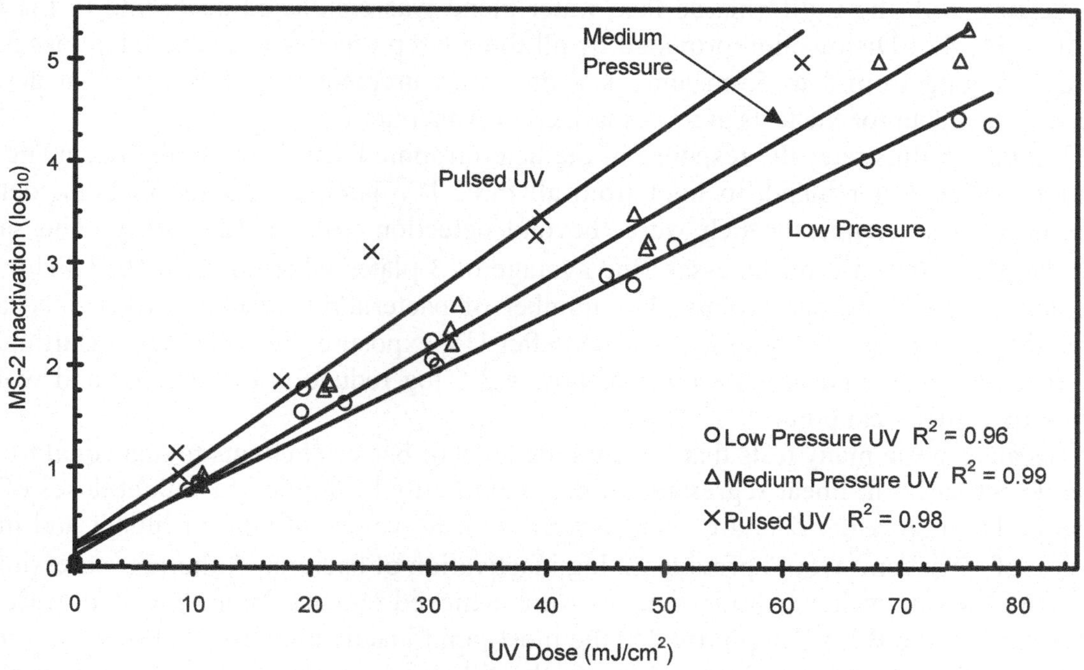

Figure 4.5 Disinfection efficiency of all three lamp types in treating male-specific coliphage MS-2 suspended in California surface water

HETEROTROPHIC BACTERIA

Following ozone and biologically active filtration, heterotrophic bacteria occur naturally in the filtered water. The assay which detects heterotrophic bacteria enumerates all aerobic bacteria which could grow in R2A agar (see method description in Chapter 2). Although many heterotrophic bacteria do not necessarily pose public health concerns, some (such as *Eschericha coli* 0157:H7) may be human pathogens. The concern for their disinfection after biological filtration, however, does not primarily rise from potential bacteriological pathogenicity. The interest in reducing heterotrophic bacteria after biological filtration is largely due to a desire to maintain distribution system water quality. During this study, levels of bacteria detected by this all-inclusive test for heterotrophs after ozone/biological filtration of the blended California surface waters ranged from 2.4 x 10^3 to 1.6 x 10^4 CFU/mL. Reduction of viable bacteria prior to distribution allows for a more easily maintained distribution system oxidant residual and reduces the possibility of bacterial regrowth in pipes.

Both disinfection and regrowth/repair potential of the bacteria were studied. Disinfection studies enumerated water samples immediately after they were exposed to varying levels of UV light. The regrowth/repair studies held UV-treated samples in both simulated uncovered-reservoir storage and distribution-system storage incubators. These incubators are described in Chapter 2.

DISINFECTION

Testing with the California surface water blend was conducted according to the matrix shown in Table 4.3. Disinfection provided by all three lamp types was evaluated across 5 target dose steps ranging from 5 to 55 mJ/cm^2. UV dose measurement was similar to that described previously. Raw data for the tests are summarized in Appendix D.

Figure 4.6 illustrates the response of the heterotrophic bacteria, naturally occurring in the California surface water blend, to light from all three UV lamps. Data shows both 'detected' samples as well as samples which were 'beyond detection' of the R2A assay. The samples beyond the detection limit of the assay (the average of 3 plates which is $\leq 2 \times 10^0$ CFU/mL) are represented by the filled data points. The number of bacteria/mL available (described above) provided the ceiling for the log-reduction seen after UV exposure. For example, a starting count of 3×10^3 CFU/mL would only be able to show a 3.5 log reduction of bacteria, and would be marked with a filled data point.

Because of the many tests that lowered the level of bacterial counts to that equal to or less than the detection limit, linear regression of data could only be applied to data at doses of 5 and 10 mJ/cm^2. The regression data are characterized with R^2 values of 0.97 or higher and indicate that a dose of 10 mJ/cm^2 would provide ≥ 2 log inactivation of bacteria. Although regression lines show somewhat similar inactivation kinetics were achieved by each lamp, the small differences may indicate that the P-UV lamp provided the most rapid inactivation/dose followed by the MP-UV lamp and then the LP-UV lamp. Because the filled data points were not considered when conducting the linear regression analysis, the regressions were seen as linear. However, because of the highly-mixed population of bacteria which are assayed in the R2A procedure, it would be difficult to understand if even statistical differences in dose-response are significant.

Table 4.3
Heterotrophic bacteria experiment matrix

Parameter	Levels	Values
Water type	1	California natural water matrix
Lamp type	3	Low-pressure, medium-pressure and pulsed
Target UV doses*	5	5, 10, 15, 35, and 55† mJ/cm^2

* A radiometer estimated exposure times before conducting experiments, all UV doses were repeated three times for each lamp.

Repair and Regrowth After Exposure to Ultraviolet Light

Table 4.4 outlines the experiment matrix followed when conducting the heterotrophic bacteria repair and regrowth testing. This testing evaluated the effects of storage on the potential regrowth or repair of heterotrophic bacteria after exposure to UV light. Samples were stored in daylight-simulated (i.e., simulating an uncovered storage reservoir) and dark-simulated (i.e., simulating an enclosed distribution system) environments for up to one week after UV exposure. The tests evaluated the effects for up to 7 days of storage in simulated daylight and simulated dark environments on UV treated samples and samples treated with UV and chloramines.

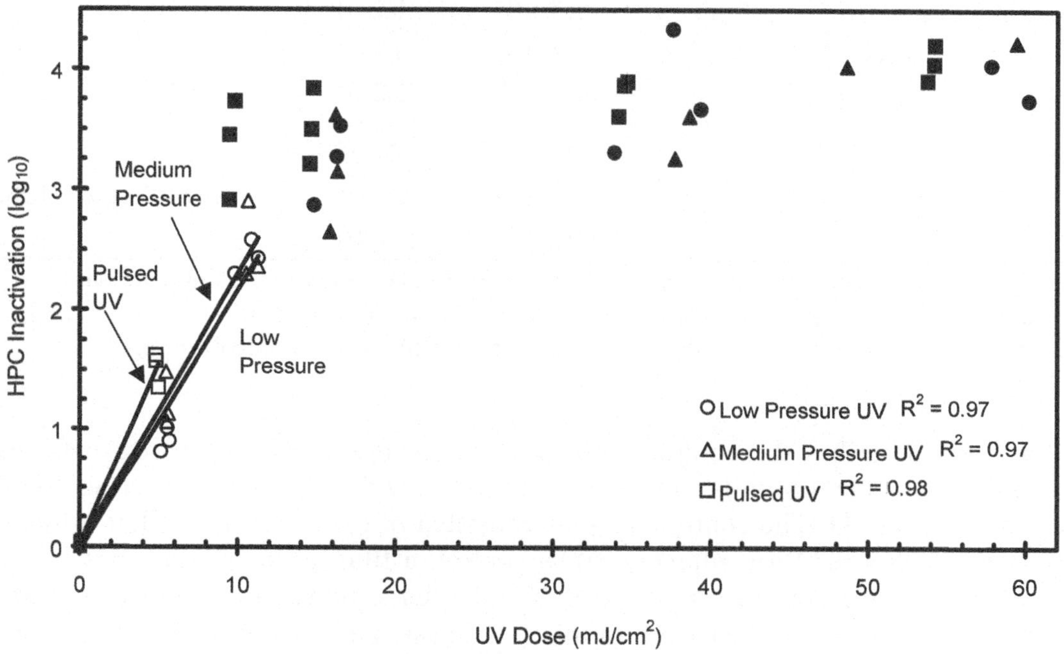

Figure 4.6 Disinfection efficiency of all three lamp types in treating California surface water containing naturally occurring heterotrophic bacteria (filled data points represent data beyond the detection limit of the R2A assay)

Table 4.4
Heterotrophic bacteria repair and regrowth experiment matrix

Parameter	Levels	Values
Water type	1	California natural water matrix
Lamp type	3	Low-pressure, medium-pressure and pulsed
Post-UV sample storage time	3	0, 3, and 7 days
Target UV doses *	5	40 and 140 mJ/cm^2 †
Target chloramine dose	1	2.5 mg/L

* A radiometer estimated exposure times before conducting experiments, all UV doses were repeated three times for each lamp.

† To be able to conduct experiments with all three lamps at once, only two doses were run at one time.

Table 4.5
Control samples collected during heterotrophic bacteria repair and regrowth testing

Control Sample Identification	Number of Storage Days (days)	Number of Bacteria in Sample (CFU/mL) Test No. 1		Test No. 2	
		Light	Dark	Light	Dark
Sterilized filter-effluent water	0	$<1\times10^0$	$<1\times10^0$ *	$<1\times10^0$	$<1\times10^0$ *
	3	$<1\times10^0$	$<1\times10^0$	$<1\times10^0$	$<1\times10^0$
	7	$<1\times10^0$	1.1×10^2	5.8×10^1	9.0×10^1
Filter-effluent water	0	1.7×10^3	1.7×10^3 *	4.0×10^3	4.0×10^3 *
	3	1.4×10^6	5.1×10^5	4.6×10^5	2.5×10^5
	7	1.1×10^6	5.2×10^5	3.8×10^5	2.2×10^5
Regrowth sample	0	1×10^0	1×10^0 *	1×10^0	1×10^0 *
	3	$<1\times10^3$ †	$<1\times10^3$ †	2.1×10^2	1.8×10^4
	7	$<1\times10^2$ †	5.6×10^3	9.2×10^5	1.7×10^4

* Control samples analyzed on day 0 were collected prior to placement in the light and dark environments, so the same value is applied to both conditions for each test

† Number of bacteria was below the detection limit set by R2A assay sample dilutions

Table 4.5 describes the control samples collected during the repair/regrowth tests and how they fared over the seven-day incubation period. Graphical representation of this data is presented in Appendix D. The control samples consisted of (1) a sterilized filter effluent water (filter sterilized with a 0.22 µm filter) negative control, a filter effluent water positive control, and a regrowth positive control sample (0.1 mL of filter effluent water added to 100 mL of sterilized filter effluent water, then 1 mL of this solution was added to the 50 mL control sample for incubation). After two tests were conducted, it was realized that there were great difficulties in providing a contamination-free environment for the incubated samples. Nearly all the sterile water negative controls contained heterotrophs by the end of the seven day test period.

Results presented below summarize the data collected during two tests when samples were incubated following UV exposure.

Uncovered-Reservoir Simulated Storage

Figure 4.7 shows the effect of storage time on the population of heterotrophic bacteria in samples exposed to (1) 40 mJ/cm^2 of UV light and (2) 40 mJ/cm^2 of UV light followed immediately by the addition of 2.5 mg/L chloramines. The UV-only sample contained nearly 10^5 CFU/mL of bacteria after three days and more than 10^6 CFU/mL of bacteria after seven days. The UV+chloramines sample contained consistently low levels of bacteria (10^0 CFU/mL) for the entire seven day period even with the chloramine residual degrading from its original value of 2.5 mg/L to a final value of 0.3 mg/L.

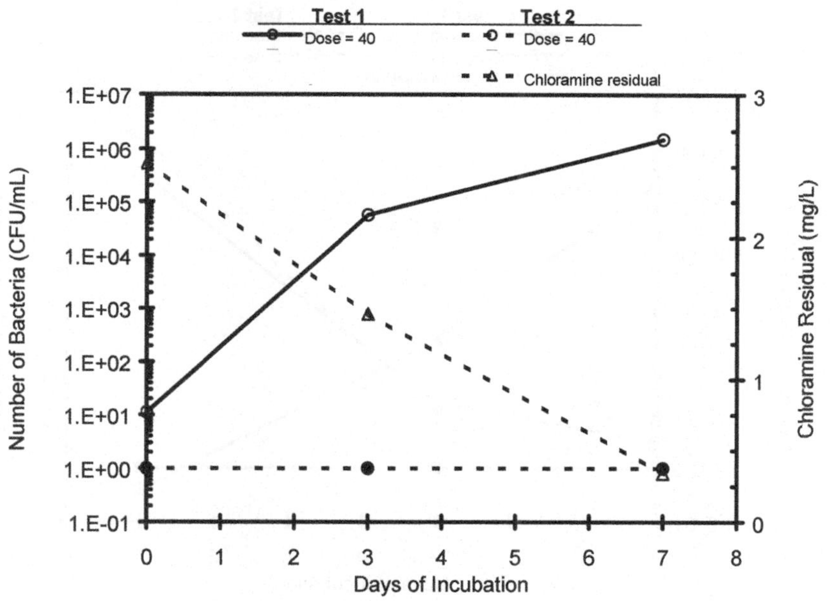

Figure 4.7 Effect of uncovered reservoir-simulated storage (light incubation) on naturally occurring heterotrophic bacteria treated by ultraviolet light from a low-pressure lamp (filled data represent samples dosed with ultraviolet light and 2.5 mg/L chloramines)

Figure 4.8 shows the effect of time on bacterial population after exposure to both 40 (with and without chloramines) and 140 mJ/cm^2 (without chloramines) of MP-UV light. Data for the 40 mJ/cm^2 dose is similar to the data produced after exposure to the LP-UV lamp, reported above. The data for the 140 mJ/cm^2 sample showed that fairly low bacterial counts could be maintained after three days post exposure, however, after seven days there were nearly 10^6 CFU/mL detected in the sample.

Figure 4.9 summarizes data collected for samples exposed to the P-UV lamp. These samples were exposed to 40 mJ/cm^2 of UV with chloramines and 140 mJ/cm^2 without chloramines. The data for the UV+chloramines samples is similar to what was reported above—bacteria remained at the 10^0 CFU/mL level for the full seven days. For the 140 mJ/cm^2 data, results were somewhat different between the two tests reported. For Test 1, there were 10^5 CFU/mL of bacteria present after seven days. For Test 2 this number was much lower, showing $<10^1$ CFU/mL after seven day incubation.

Due to difficulties in continuing testing (the biological filters were chlorinated due to treatment plant operational needs), the data sets were not consistent across lamp-type and were not repeated more than twice. However, the limited amount of data provided evidence of certain trends. After application of UV alone and incubation for seven days, it was most likely that bacterial counts would return to levels that were seen in the positive control samples (filter effluent sample with no treatment). When UV treatment was combined with chloramines, results clearly showed that even though the chloramine residual degraded significantly, application of a UV dose of either 40 or 140 mJ/cm^2 followed by chloramines provided $<2\times10^0$ CFU/mL of bacteria after seven days. This clearly indicates that water treated by UV light would not have biological stability to remain bacteria free if secondary treatment was not administered. More data would need to be collected to understand if high doses of P-UV light, without chloramine addition, could consistently provide low bacteria counts in a daylight environment.

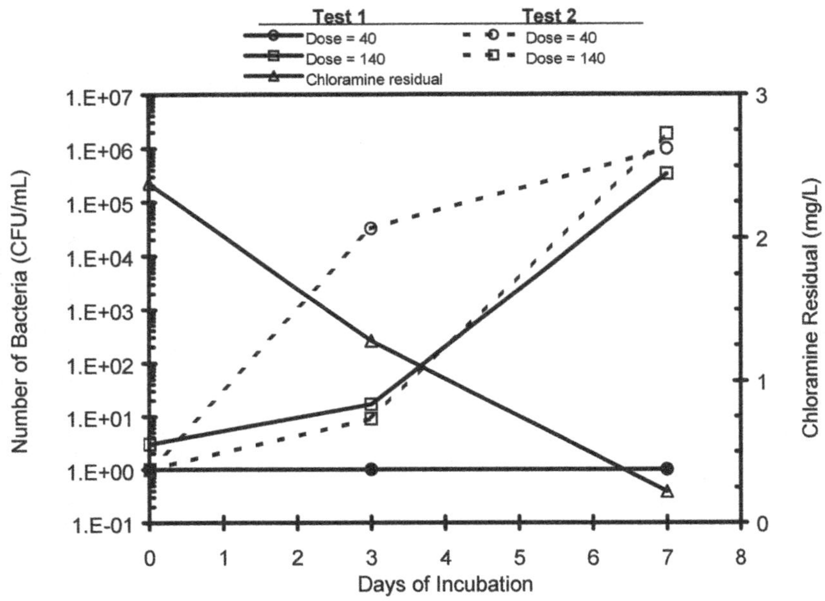

Figure 4.8 Effect of uncovered reservoir-simulated storage (light incubation) on naturally occurring heterotrophic bacteria treated by ultraviolet light from a medium-pressure lamp (filled data represent samples dosed with ultraviolet light and 2.5 mg/L chloramines)

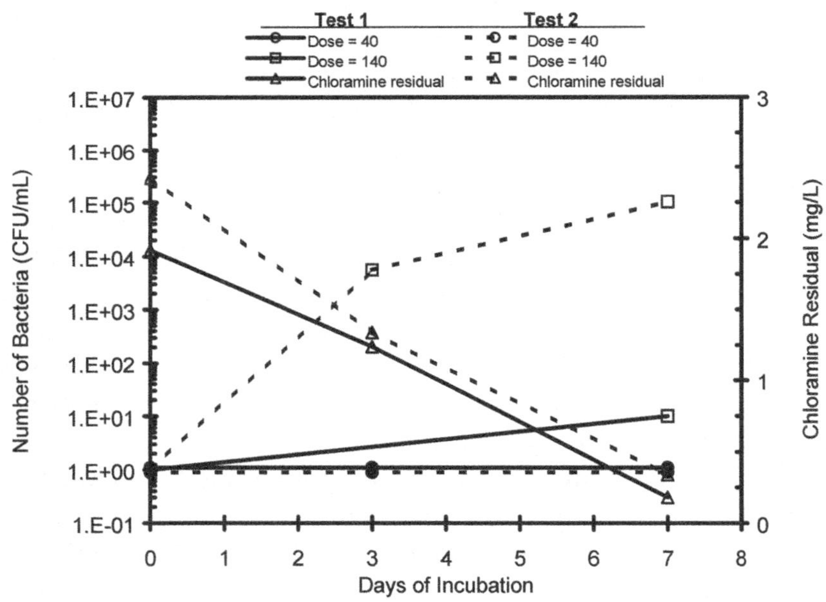

Figure 4.9 Effect of uncovered reservoir-simulated storage (light incubation) on naturally occurring heterotrophic bacteria treated by ultraviolet light from a pulsed lamp (filled data represent samples dosed with ultraviolet light and 2.5 mg/L chloramines)

Distribution-System Simulated Storage

Figures 4.10, 4.11, and 4.12 summarize results for the distribution-system storage simulations after samples were exposed to LP-, MP-, and P-UV light, respectively. These data were collected in parallel with the light-simulation tests, so the same doses per lamp were incubated in the dark as were incubated in the lighted environment described in Chapter 2.

It was observed that the data for the samples exposed to the low UV dose of 40 mJ/cm^2 and incubated in the dark responded similarly to the samples incubated in the light. After three days there were approximately 10^5 CFU/mL of bacteria present and even more were present after the seven-day incubation. Also, the dose of 40 mJ/cm^2, when followed by 2.5 mg/L of chloramines (reduced to approximately 0.7 mg/L after seven days), maintained a very low count of bacteria after the seven day incubation for all three lamp datasets.

For the dark-incubation tests, the number of bacteria present after the high UV dose of 140 mJ/cm^2 was slightly different than for the light-incubation tests. The MP-UV data indicated relatively low counts of bacteria in one of the two tests, while the P-UV data indicated high counts after both test samples were incubated seven days.

The data from both incubation environments indicate that biological stability can be maintained if the UV-exposed samples were coupled with the addition of chloramines. Even with the degradation of the chloramine residual over the seven-day incubation periods, all UV+chloramine samples showed very low counts of bacteria. Without the chloramine residual, and even with a nearly 4-fold increase in the UV dose to 140 mJ/cm^2, the bacteriological quality of the water was consistently compromised. This compromise in biological stability caused levels of heterotrophic bacteria to return to levels similar to that of untreated, biological-filter effluent.

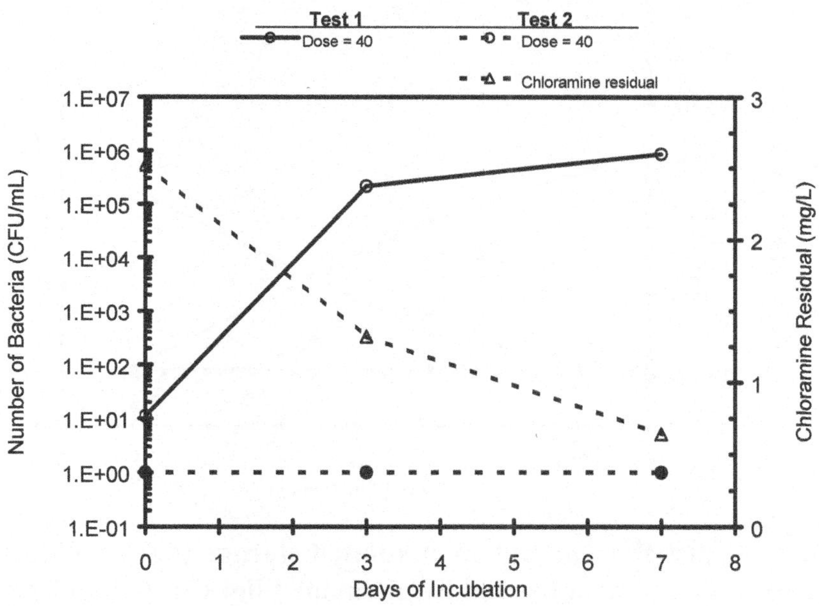

Figure 4.10 Effect of distribution-system simulated storage (dark incubation) on naturally occurring heterotrophic bacteria treated by ultraviolet light from a low-pressure lamp (filled data represent samples dosed with ultraviolet light and 2.5 mg/L chloramines)

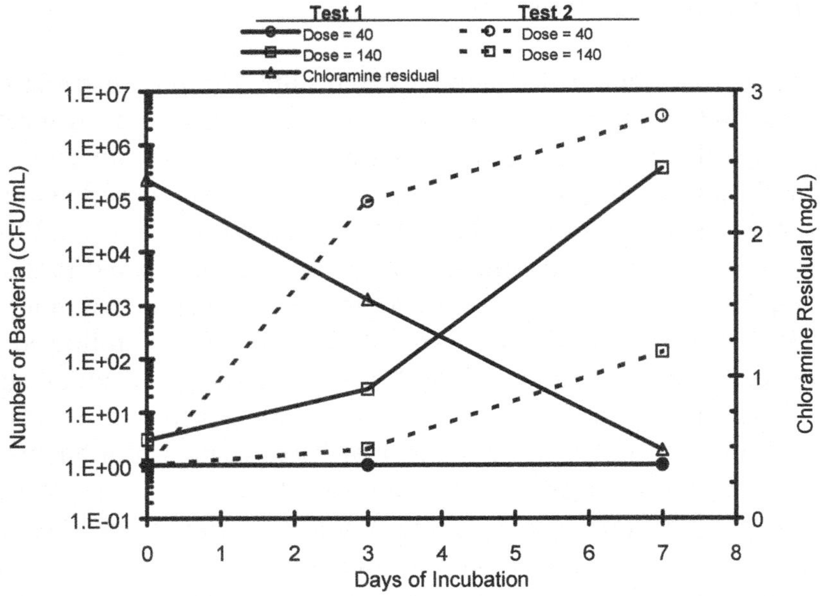

Figure 4.11 Effect of distribution-system simulated storage (dark incubation) on naturally occurring heterotrophic bacteria treated by ultraviolet light from a medium-pressure lamp (filled data represent samples dosed with ultraviolet light and 2.5 mg/L chloramines)

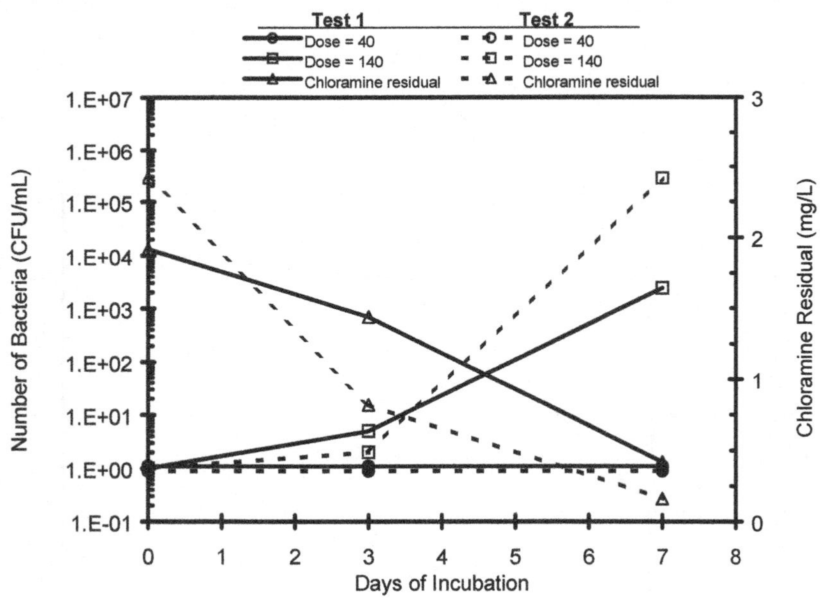

Figure 4.12 Effect of distribution system-simulated storage (dark incubation) on naturally occurring heterotrophic bacteria treated by ultraviolet light from a pulsed lamp (filled data represent samples dosed with ultraviolet light and 2.5 mg/L chloramines)

This study utilized regrowth controls (a sterilized water sample inoculated with a very small amount of bacteria from the filter-effluent water) summarized in Table 4.5. These controls were maintained to attempt identification of bacterial regrowth and repair. The regrowth controls showed significant amount of regrowth of bacteria over the seven day period in both lighted and dark environments, although the absolute levels of regrowth varied (from non-detect to 9 x 105). Thus, because this control protocol is also representative of the conditions under which bacteria experienced after UV exposure, it can be hypothesized that the recovery in heterotrophic bacteria levels following UV exposure is at least in part due to regrowth of surviving bacteria. However, given the limited data set and the variability seen in the experiment controls, it is difficult to differentiate between regrowing bacteria and those that repaired themselves after UV damage.

CRYPTOSPORIDIUM PARVUM

C. parvum was suspended in the pretreated California surface water blend and exposed to varying levels of UV light according to the testing matrix shown in Table 4.6. As described in Chapter 2, the oocysts were assayed by human cell culture infectivity assay after exposure to UV light to determine the effect UV light had on the infective potential of the oocysts.

Logit Response of Human Cell Culture Assay

Figure 4.13 shows the resulting logit response of the cell culture assay. The data for the logit model were calculated using the method of least-squares as described by Finch et al. (1994). The resulting model relationship between the density of oocysts introduced to the cell monolayer vs. response logit was calculated to be:

$$\text{Logit} = -2.06 + 1.09\ \log_{10}(\text{oocysts added to cell monolayer}) \qquad (4.1)$$

Based on this data, the infectious dose where 50 percent of the individual cell monolayers were infected was 1.9 log, or 77 oocysts. This logit response is a cumulation of all the different *C. parvum* lots tested during this study.

Table 4.6
***Cryptosporidium parvum* experiment matrix**

Parameter	Levels	Values
Water type	1	Ozonated/biofiltered natural water matrix
Lamp type	3	Low-pressure, medium-pressure and pulsed
Target UV doses*	5	1, 4, 7, 11, and 25 mJ/cm^2

* Radiometer estimated exposure times before conducting experiments; all UV doses were repeated three times for each lamp.

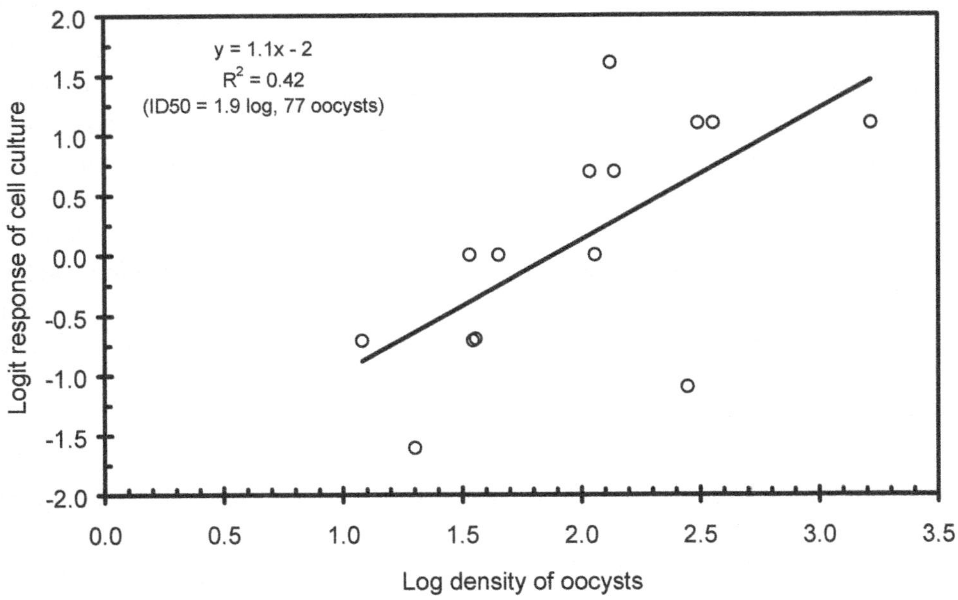

Figure 4.13 Logit response of human cell culture infectivity assay to the application of increasing numbers of oocysts of *C. parvum*

Effect of Ultraviolet Light on *Cryptosporidium parvum*

Table 4.7 summarizes the positive control (zero dose) samples. These were contained in petri dishes and placed under the collimated beam (while light from the lamp was blocked) for the duration corresponding to the maximum UV exposure time. Positive controls were given a cumulative UV dose of 0 mJ/cm^2. Due to cell culture limitations, the exact number of infective oocysts in the positive controls (zero dose sample) could not be precisely calculated. This was due to over-bleaching of the gel images quantified by the digitization/gel quantitation procedure used. This tabulated summary also shows how the recorded 'infective' number of oocysts compared with hemocytometer and IFA counts. On average, the actual number of infective oocysts was 7 percent of the hemocytometer count and 15 percent of the IFA counts. Although cysts were consistently new and exhibiting high excystation rates (see discussion in Chapter 2), fairly low numbers of oocysts were actually seen as infective.

The information plotted in Figure 4.14 includes the positive controls described above, the quantifiable UV-dose, oocyst response data for exposures to all three UV lamps, and unquantifiable data (data beyond cell culture assay detection limits). Appendix D tabulates the data for these experiments. Tests were conducted over a seven month period with two UV doses per lamp, including controls, in each of nine separate experiments. The UV-dose, oocyst response data within the detection limits of the infectivity assay indicate significant reduction in infective oocysts with increasing UV dose for all UV lamps. These data are similar to the data presented by other researchers in the literature using infectivity as a disinfection end-point. The US EPA will be summarizing all the *Cryptosporidium* UV inactivation data in the process of rulemaking for the Long Term 2 Enhanced Surface Water Treatment Rule which will detail

Table 4.7
Summary of positive control samples for *Cryptosporidium parvum* experiments

Test Number	Hemocytometer Count * (No)	IFA Count (No.)	Scanned Gel Image Count † (N)
1	1×10^5	N/A	2×10^4
2	1×10^5	N/A	9×10^3
3	1×10^5	N/A	9×10^3
4	1×10^5	N/A	4×10^3
5	3×10^5	N/A	7×10^3
6	1×10^5	2×10^4	7×10^3
7	1×10^5	2×10^4	8×10^3
8	1×10^5	7×10^4	6×10^3
9	1×10^5	8×10^4	8×10^3

N/A = not analyzed

* Hemocytometer count was conducted on the stock oocyst solution, then this was diluted to provide the number of oocysts shown in the positive control

† Scanned gel image values may be greater than or equal to values shown here

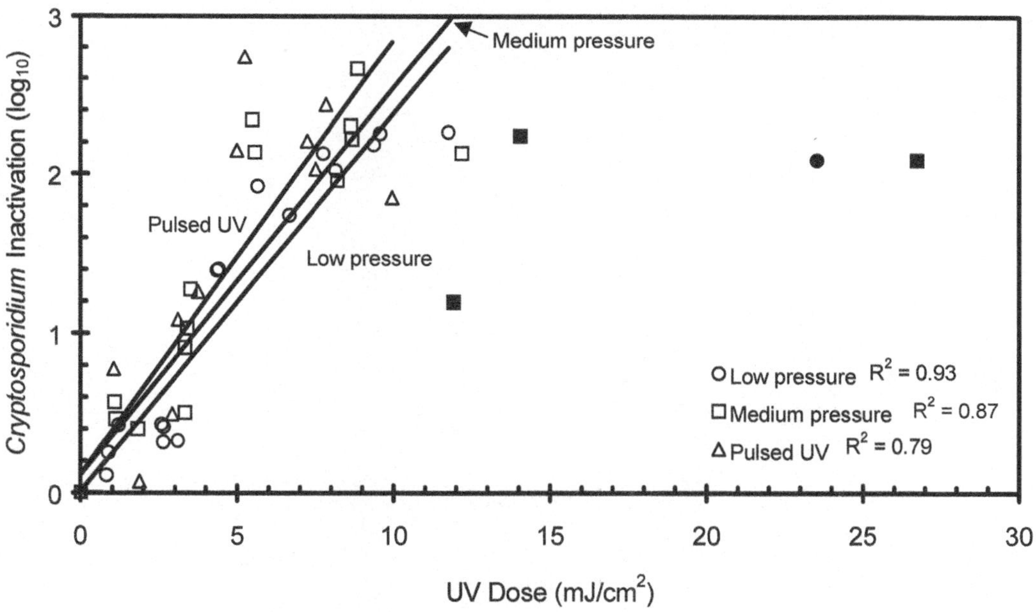

Figure 4.14 Disinfection efficiency of all three lamp types in treating *Cryptosporidium parvum* suspended in California surface water (filled data points represent data beyond the detection limits of the cell culture infectivity assay)

disinfection requirements for *C. parvum*. After regression analysis, the data for the three lamps appear to provide only slightly different slopes. The regression equations were $y = 1.06\ e^{-0.56x}$, $y = 0.75\ e^{-0.56x}$, and $y = 0.77\ e^{-0.63x}$ for the LP-, MP-, and P-UV lamps, respectively. Thus for *C. parvum*, it appears that the efficacy of UV disinfection is not appreciably different for the different UV sources tested. This conclusion needs to be assessed with respect to the dose measurement methodology and the action spectra of *C. parvum*, similar to the discussion for the dose response of MS2 coliphage above. UV dose was measured using the bias of the "G" filter on the radiometer (See Figure 3.4), modeled after the action spectra of *E. coli*. As it turns out, the action spectra for *C. parvum* as identified by Linden et al. (2001) is very similar to the action spectra of *E. coli* as exemplified in the weighting provided by the G filter radiometer configuration. Thus, if the irradiance from a polychromatic, UV emission spectra was measured using the G filter, the calculated delivered UV dose should be representative of the actual effective germicidal dose for *C. parvum*, as the weighting of the G filter is similar to the action spectra for *C. parvum*. We see this holding true in the similarity of the dose response for *C. parvum* upon exposure to different UV sources as illustrated in Figure 4.14. Thus, if dose is measured using proper germicidal corrections for a specific microorganism under examination, there is no apparent difference in efficacy among the lamp types. Thus, for *C. parvum*, it appears that UV lamp type does not benefit inactivation; what is important are the methods used for dose measurement and proper documentation of those methods such that the relative efficacy of different lamps can be properly assessed.

CHAPTER 5
CONCLUSIONS AND RECOMMENDATIONS

CONCLUSIONS

A number of conclusions may be drawn from this investigation of the disinfection efficiency provided by monochromatic, and polychromatic, UV lamps. The three lamps investigated represent the range of UV lamp choices available to the water industry. Although LP, high-output lamps were not investigated in this study, it may be expected that dose-measurement protocols would mirror that for LP lamps and resulting disinfection kinetics would be similar to that of LP lamps.

Conclusions reached as a result of this study include the following:

1. The relative contribution of the germicidal UV light emitted from the P-UV lamp (200 to 300 nm range) was small relative to the entire output spectra (200 to 800 nm range).
2. UV irradiation from different UV lamp sources is not equally germicidal.
3. Petri factors vary greatly for different collimating tube length/lamp combinations, and must therefore be determined and optimized for proper UV dose measurement.
4. Radiometer and actinometer (KI and uridine) methods for measuring UV irradiance were both able to produce similar results for the LP-UV lamp.
5. Because of the polychromatic nature of the MP-UV lamp, a mathematical correction was necessary to correctly interpret the incident irradiance measured by the radiometer and the actinometers. After mathematical correction, results were similar for both measurement tools.
6. DNA absorbance spectra may be an appropriate, although not ideal, weighting factor to estimate germicidal UV irradiance. To increase accuracy of germicidal irradiance estimates, it would be preferable to use an organism-specific weighting factor.
7. A properly calibrated radiometer may be used to measure the germicidal UV irradiance from LP- or MP-UV lamps. Although a radiometer will not measure UV irradiance emitted by a P-UV lamp, it can be reliably used to determine the cumulative UV dose given to a petri-dish sample (in a collimated-beam mode) if it is equipped with an integrating function.
8. There were large differences in irradiance measurement for the P-UV lamp when using either the radiometer or an actinometric method, even after mathematical correction of its polychromatic output.
9. When using the uridine actinometer to measure very-high UV doses (i.e., >300 mJ/cm^2), there was evidence that non-detectable photobyproducts may be formed when exposed to either (1) long-duration, low- to moderate-intensity UV light or (2) high-intensity, short-duration pulses of UV light. However, if applying disinfection-level dosages of UV light from either the LP-, MP-, or P-UV lamps, the uridine actinometer solution worked adequately in predicting UV irradiance.
10. The uridine actinometer is very stable at up to 7-day storage at room temperature, although it is recommended to be stored at 4 °C when not in use and analyzed immediately after exposure to UV light.

11. The iodide/iodate actinometer is unstable, indicating that it should be made fresh on the day of use and stored in a cold, dark environment (e.g., in a 4 °C refrigerator in amber-tinted glassware), control (unexposed) samples should accompany UV-exposed samples, and the user needs to be cautious to avoid significant error in measurement of lower UV irradiation doses.
12. Comparing radiometry and actinometry with biological dose measurement (biodosimetry) techniques using *B. subtilis* spores, predicted irradiance values all fell within statistical similarity, in similtude with biodosimetric values.
13. Disinfection kinetics of *B. subtilis* were shown to be similar for all three UV lamps. The response below 20 mJ/cm^2 was significantly more rapid than to higher dosages of UV. A 4-log inactivation of spores could be achieved at 23 mJ/cm^2 while only 1-log greater inactivation could be achieved with a 4-fold increase in UV dose.
14. After water quality parameters were accounted for, similar disinfection efficiency was observed for *B. subtilis* when it was suspended in different natural water quality matrices.
15. The disinfection kinetics for MS-2 coliphage were shown to be slightly different from the three lamps. To achieve a 2-log reduction, UV doses of 32, 28, and 22 mJ/cm^2 were necessary from the LP-, MP-, and P-UV lamps, respectively. This difference may be explained by the different emission spectra from the lamps and the dose measurement methodology. This illustrates the need to develop a universal standard for measurement of UV dose provided by polychromatic lamps to avoid wavelength-specific anomalies for microbes which are non-conforming to the absorbance of DNA.
16. UV light from all three lamps provided rapid disinfection of heterotrophic bacteria produced after ozone/biological filtration. A dose of 10 mJ/cm^2 was observed to provide 2 log inactivation of bacteria. Any differences in the dose-response provided by the three lamps would be difficult to interpret due to the highly-mixed population of bacteria assayed by the R2A assay.
17. Even after treatment of up to 140 mJ/cm^2 of UV light, heterotrophic bacteria readily regrew/repaired themselves if they were not treated with a secondary, residual disinfectant. Applying 2.5 mg/L of combined chlorine (chloramines) to the samples after UV irradiation allowed bacterial counts to remain very low for seven-day incubation periods in lighted (uncovered-reservoir simulated storage) and dark (distribution-system simulated storage) environments.
18. For all three lamps tested, the disinfection kinetics of *C. parvum* were observed to be similar. After interpreting data with linear regression analyses, at least 2-log inactivation of infectious *C. parvum* was observed at a dose of 10 mJ/cm^2.
19. Combined results from the testing reported here indicate that the protozoan organism *C. parvum* was more susceptible to UV light than the other organisms tested. Heterotrophic bacteria and *B. subtilis* aerobic spores were slightly more resistant to UV light than *C. parvum* while the virus surrogate MS-2 coliphage was the most resistant.

RECOMMENDATIONS

Results presented throughout this report have provided insight into future needs for research of UV dose measurement and disinfection effectiveness. This section provides guidance and recommendations to scientists and utilities that may be interested in either better understanding UV-dose measurement techniques or implementing UV technology for disinfection of water. Recommendations include:

1. Before using chemical actinometry, a full understanding of actinometer-specific inadequacies (i.e., reaction to visible light, stability issues, effectiveness in absorbing germicidal UV light) is required.
2. Due to the polychromatic and high-intensity output of many new UV lamp technologies, special considerations may be necessary to allow for proper actinometer and/or radiometer measurement of UV irradiance.
3. If using a radiometer to measure UV irradiance from any source, make sure that it is properly maintained and checked against standards and/or calibrated by the manufacturer on no less than a yearly basis.
4. Biological actinometry (biodosimetry) can be an effective tool to determine the germicidal effectiveness of a lamp that emits UV light. To properly use this tool, the action spectra of the microorganism used (how it responds to different wavelengths of UV light) must be understood and considered when interpreting test results for polychromatic light.
5. It is important for future research to investigate the use of different dose-weighting schemes and analyze dose-response of microorganisms using their specific germicidally-effective wavelengths.
6. When comparing the lamps in this study on an equivalent UV-dose measurement basis, equivalent disinfection was observed for *C. parvum* and *B. subtilis* no matter the lamp type used. For MS-2 coliphage, there appeared to be greater inactivation under the polychromatic systems when UV doses were based on a DNA absorbance weighting, thus the findings may be an artifact of the DNA based weighting system. Regardless, utilities may have site-specific issues that indicate benefits of using one lamp technology over another. These issues may include flow rate (large flow rates may require greater numbers of lamps) and water quality issues (possible fouling of high-temperature lamps).
7. Before installing a UV system, a utility should understand what organisms are targeted for disinfection by the process because different target organisms may have order-of-magnitude difference in required UV dose for equivalent disinfection.

Practical Considerations From the Research

For utilities considering the use of UV in their treatment facilities, there are a number of important outcomes from this research that need to be recognized. When choosing between different UV light sources, it is apparent that there is no overwhelmingly conclusive difference in the efficacy of one lamp type over another for the microorganisms tested. What is important is that care and discretion is taken in evaluating UV disinfection data used to make a decision on choice of UV technology. Ultimately, the differences in germicidal efficiency will be one part of a larger matrix from which the disinfectant choice will be made. It should also be clear from the findings that measurement of UV dose is not as straightforward as is desirable. Although not addressed in this report, these issues add to the complexity of trying to determine UV dose on a reactor scale where hydraulics are also a factor. Therefore, validation of successful disinfection is necessary to complement any claims of dose delivery in a given UV reactor.

This research shows that not all microorganisms are equally affected by UV light and that some wavelengths may be more effective for inactivation than others for specific microbes. However, as more wavelengths are available, the means of interpreting the germicidal intensity of the UV source become more complex. Tools such as actinometry (evaluated here) and mathematical modeling are promising for unraveling the complexities of polychromatic UV dose measurement.

The use of a biodosimetry test, such as would be used during a reactor validation, to measure the delivered dose for a polychromatic UV lamp (e.g. MP-UV) shows good agreement with actinometry and radiometry dose measurement methods. These results indicate that biodosimetry performed on a UV reactor containing polychromatic lamps is an appropriate, and practical, means of determining dose in such systems. Inactivation of heterotrophic bacteria by UV is possible but subsequent to inactivation, these microbes are likely to regrow without addition of a secondary residual disinfectant. These findings, even for high UV doses, support the need for utilities to develop robust treatment strategies that use multiple disinfectants to achieve inactivation of pathogens and nuisance bacteria. The results also support recent findings that UV is an excellent disinfectant for inactivation of *Cryptosporidium* and will support the USEPA decision making for determining the appropriate UV dose for different levels of *Cryptosporidium* inactivation.

APPENDIX A
SELECTED STUDY PHOTOGRAPHS

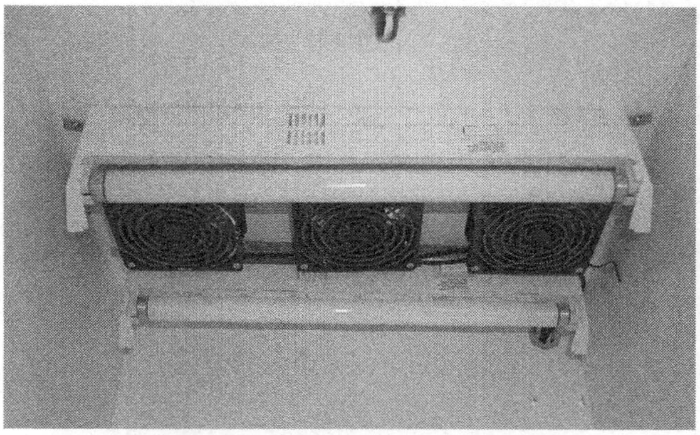

Inside, top of light incubator showing the lamps and exhaust fans (air exit)

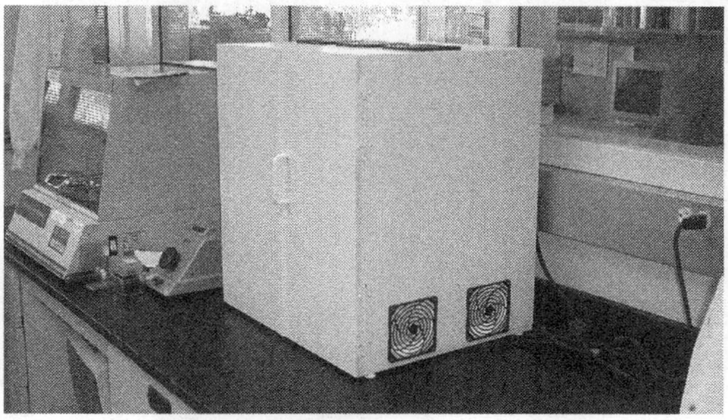

Side, exterior view of light incubator showing front doors and side vents (air inlet)

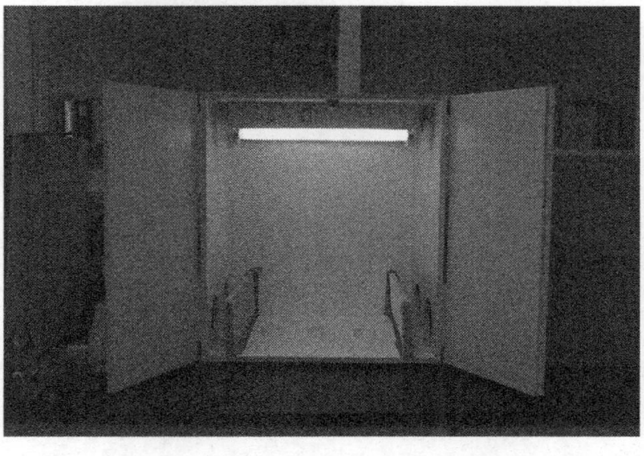

Front, exterior view of light incubator showing sample location and baffles used to decrease air flow across the petri-dish samples

Daylight-simulated incubation box used for heterotrophic bacteria samples

Low-pressure and Medium-pressure collimated-beam apparatus and power supply (center) and International light radiometer and detector (lower right)

Pulsed-ultraviolet, collimated-beam apparatus (center) showing the power supply (right) and black control panel (lower left)

APPENDIX B
URIDINE PHOTOHYDRATE BYPRODUCT ANALYSIS

INTRODUCTION

A Micromass AutoSpec-oaTOF mass spectrometer (MS) was used in the electrospray positive ion mode with a magnetic sector analyzer. Samples were delivered from a syringe pump at a flow rate of 50 μL/min. The solvent used was 1:1 MeOH:$H_2$0 (methanol). Sample spectra obtained were compared with that of the solvent blank to subtract the background peaks.

MS/MS analysis of the uridine sodium adduct ion with a mass-to-unit charge ratio (m/z) of 267 was conducted using time of flight (TOF) as the second mass analyzer. Both methane and xenon were tested as collision gases to provide different collision energies. In both cases, Na^+ appeared to be the main fragmentation peak. This is consistent with the $[uridine+Na]^+$ parent ion. Quantitiative analysis of uridine and the resulting photohydrate using MS was made possible by using selected ion recording (SIR) under voltage scan mode. Samples were introduced into the electrospray (ES) ionization source through a loop injector with a 10 μL loop. The flow rate of the mobile phase was 50 μL/min. The instrument was calibrated with $PEGH^+$ using PEG 200. The peaks corresponding to each specific ion were integrated to give the peak areas, which were used in all the calculations involved.

In order to determine whether there were other photodecomposition products formed upon exposure to UV light, a preliminary liquid chromatography/ES interface/MS (LC/ESI/MS) experiment was performed. A Waters Nova-Pak C18 column (2 mm x 150 mm) was used and the gradient (described in Table B.1) was employed at a flow rate of 100 μL/min.

SOLVENT SELECTION

Three different solvent types were examined using a uridine standard to achieve the best sensitivity. When 1:1 CH_3CN:H2O was used as the solvent, $[2M+Na]^+$ (m/z 511) was present at low levels. However, when the solvent was changed to methanol, the monomer $[M+Na]^+$ (m/z 267), the dimer $[2M+Na]^+$ (m/z 511), and the trimer $[3M+Na]^+$ (m/z 755) were detected with very good sensitivity. The monomer recorded the strongest peak followed by the dimer and then the trimer. When 5 percent acetic acid was added to methanol to enhance the ionization, both $[M+H_2O+H]^+$ and $[M+H]^+$ were detectable from the spectrum. Since the uridine photohydrate has been previously reported as sensitive to acidic and high-temperature conditions (Sinsheimer and Hastings 1949, Swenson and Setlow 1963), the solvent methanol was used in all experiments conducted.

Uridine Extinction Coefficient

UV absorbance of three uridine solutions in Milli-Q® water was measured at 260 nm as shown in Table B.2. The path length of the UV cell was 1 cm. The extinction coefficient was determined to be 9.68 $x10^3$ by taking the average of the three readings.

Table B.1
Solvent gradient used when analyzing for uridine photodecomposition products

Time	Percent H_2O	Percent MeOH	Function
0	90	10	
1	90	10	hold
6	5	95	linear
20	5	95	hold
30	5	10	linear

Table B.2
Ultraviolet absorbance at 260 nm

Concentration (mM)	Absorbance	Extinction Coefficient (L/mol•cm)
0.057	0.540	9.47×10^3
0.103	1.005	9.76×10^3
0.202	1.979	9.80×10^3

EXPERIMENT CONDITIONS

To tune the instrument for best sensitivity, uridine was used as a standard and its peak of $[M+Na]^+$ (m/z 267) was maximized. The cone voltage was set at 6.4 v, the needle voltage at 28.0 v and the ring electrode at 17.6 v. These tune parameters were constant throughout the project.

STANDARD SELECTION

First, glycine-phenylalanine (Gly-Phe) which has a strong $[M+Na]^+$ peak at m/z 245 was used as a standard. A linear calibration curve of uridine was obtained when plotting the ratio of the peak area of uridine divided by the peak area of Gly-Phe versus the concentration of uridine. However, in the case of irradiated uridine with the presence of the photohydrate, Gly-Phe showed abnormal peak shape and intensity, suggesting possible reaction between the Gly-Phe and the photohydrate.

The next choice for a standard was caffeine, which gave a strong $[M+Na]^+$ peak at m/z 217. Because of its stable structure, it was not expected to have any chemical interactions with either uridine or the photohydrate. To study the interaction of the caffeine with the uridine, 50 μL of 0.52 millimolar (mM) caffeine stock solution dissolved in methanol was added to 1 mL of uridine solution of ranging concentrations (from 0.006 to 0.075 mM). The peaks corresponding to caffeine (m/z 217) and uridine (m/z 267) were acquired simultaneously and integrated to give the corresponding peak areas. The calibration curve of uridine was obtained in a similar fashion to what was attempted with the Gly-Phe, and is shown in Figure B.1. The average of two runs gave the equation B.1:

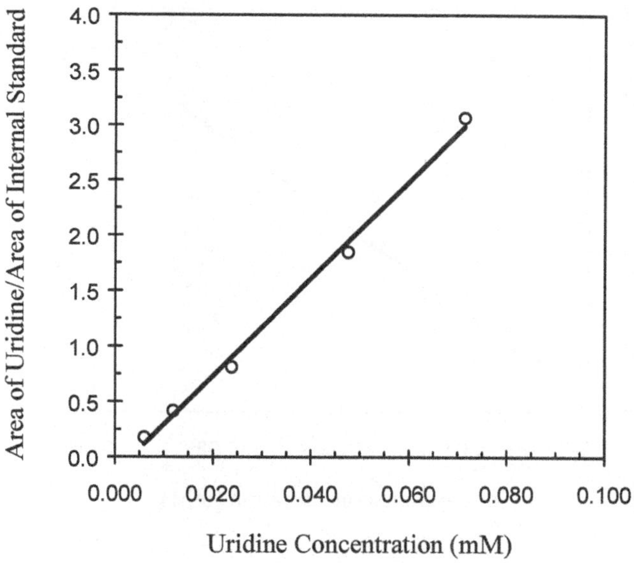

Figure B.1 Calibration curve of uridine (caffeine standard)

$$\frac{\text{Area of uridine}}{\text{Area of caffeine}} = 41.7 \times [\text{uridine}] - 0.081 \quad \text{(B.1)}$$

where [uridine] = concentration of uridine

In the presence of uridine, the sensitivity of the caffeine peak was suppressed to a small extent. For example, the higher the uridine concentration, the higher the extent of caffeine suppression. Attempts to overcome this suppression were made by adding additional Na^+ ions (0.05 mM NaCl) to provide enough Na^+ ions for both caffeine and uridine, but were not successful. The source of this suppression remains unknown.

The calibration curve of the photohydrate was constructed with the following multi-step process. First, 1 ml of 0.1 mM uridine solution irradiated at 3,000 mJ/cm^2 with the medium-pressure UV lamp. This solution was then diluted 2-, 4-, and 8-fold with methanol and 50 μL of 0.52 mM caffeine stock solution was added. Next, the $[M+Na]^+$ ions corresponding to caffeine, uridine, and the photohydrate at m/z 217, 267, and 285, respectively, were acquired simultaneously and integrated to give the corresponding peak areas. After the above, the concentrations of uridine in each sample were calculated from dividing the peak area of uridine by the peak area of caffeine standard according to equation B.1. Then, assuming the photohydrate was the only irradiation product of uridine, the concentration of the photohydrate was estimated from the following equation in the initial solution:

$$[\text{Photohydrate}]_i = [\text{Uridine}]_0 - [\text{Uridine}]_i \quad \text{(B.2)}$$

where $[\text{Uridine}]_0$ = starting concentration of uridine
$[\text{Uridine}]_i$ = concentration of uridine after a given dose
$[\text{photohydrate}]_i$ = concentration of photohydrate produced after a given dose

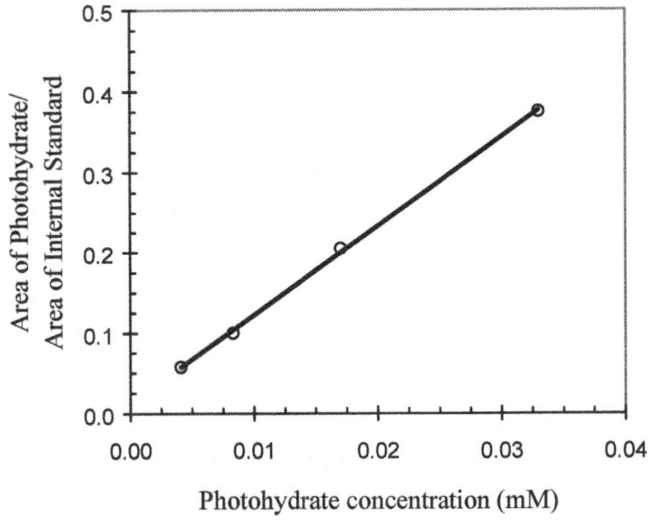

Figure B.2 Calibration curve of the photohydrate

The concentrations of the photohydrate in the diluted solutions were calculated by dividing the concentration of the photohydrate in the initial solution by the appropriate dilution factor.

Finally the area of photohydrate divided by the area of caffeine standard versus the concentration of the photohydrate was plotted to provide the calibration curve shown in Figure B.2. The average of two runs gave the following equation:

$$\frac{\text{Area of Photohydrate}}{\text{Area of Internal Standard}} = 11.8 \text{ x } [\text{Photohydrate}]_i + 0.005 \tag{B.3}$$

Five validation samples of uridine solution were exposed to doses of 0, 30, 100, 300, and 1,000 mJ/cm^2 (UV irradiance measured by radiometer) from a low-pressure UV lamp at Duke University and analyzed using the SIR voltage scan method described above. The $[M+Na]^+$ ions corresponding to caffeine, uridine, and the photohydrate (at m/z 217, 267, and 285, respectively) were acquired simultaneously and integrated to give the corresponding peak areas. The concentrations of uridine and the photohydrate were calculated using equations 1 and 3, respectively. The concentrations of uridine and the photohydrate in each sample were determined from the average of four runs. These concentrations were multiplied by a factor of 2 because they were diluted in half with methanol when processed in the laboratory. The final concentrations of uridine and the photohydrate in this set of validation samples are tabulated in Table B.3 and the uridine concentration versus UV dose curve is shown in Figure B.3.

Results shown in Figure B.3 illustrate that by using the MS method, it is possible to quantify the uridine photohydrate generated from exposure to UV light. This is an advantage of MS over UV-VIS spectrometry. This increase of the photohydrate was shown to have a linear correlation with UV dose, as opposed to the degradation of uridine which was a second-order relationship. Furthermore, the sum of $[\text{uridine}]_i$ and $[\text{photohydrate}]_i$ at different UV doses equaled $[\text{uridine}]_0$, which validated the quantification methodology.

Table B.3
Concentration of uridine and photohydrate in validation samples

UV Dose (mJ/cm^2)	Concentration of Uridine (mM)	Concentration of Photohydrate (mM)
0	0.0502	0.0004
30	0.0470	0.0008
100	0.0432	0.0024
300	0.0434	0.0066
1000	0.0352	0.0216

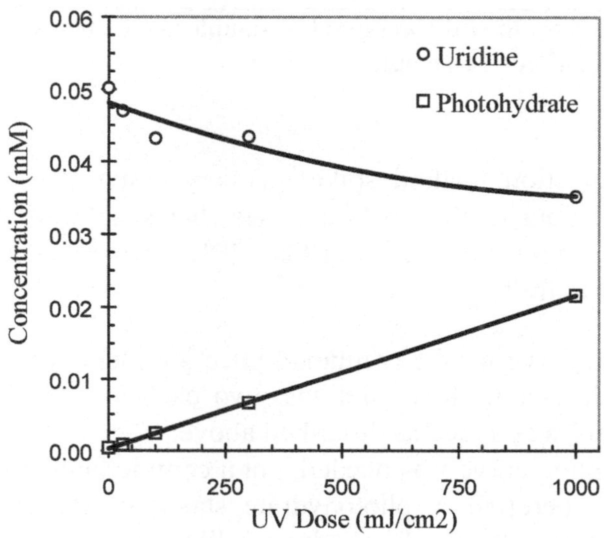

Figure B.3 Uridine and photohydrate relationship to UV dose for uridine samples exposed to light from a low pressure lamp

SPIKE RECOVERY AND CALIBRATION CURVES

There were two methods evaluated for introducing the samples and caffeine standard to the MS. The methods evaluated were (1) combining the caffeine standard with the samples and (2) injecting the caffeine standard in an alternating fashion with the samples. Table B.4 summarizes the spike recoveries when using both of these methods.

For the combined injection method, 2 μL of 0.737 mM uridine stock solution in methanol was spiked into 1 mL of each irradiated uridine sample to give a spike concentration of 0.0015 mM. Then 2 μL of 1.3 mM caffeine stock solution in methanol was added to give a final caffeine concentration of 0.0026 mM. Spike recoveries for this method, although less than 100 percent, fell within acceptable QA/QC limits. This indicates that there are possible negative biases for the method.

Table B.4
Caffeine standard spike recoveries calculated from experiments

Sample Identification	Spike Recovery for Combined Injections	Spike Recovery for Alternating Injections
Uridine Blank Solution A*	71	54
Low-Pressure UV Sample 1	91	74
Low-Pressure UV Sample 2	75	67
Uridine Blank Solution B*	75	NA
Medium-Pressure UV Sample 1	70	45
Medium-Pressure UV Sample 2	81	84
Pulsed-UV Sample 1	NA	34
Pulsed-UV Sample 2	NA	105

* Blank solution A was prepared for the low-pressure UV samples while blank solution B was prepared for the medium-pressure and pulsed-UV samples.

For the alternating injection method spike recovery results were somewhat different (results shown in Table B.4). Four of the seven samples (blank solution A, LP-UV sample 2, MP-UV sample 1 and P-UV sample 1) gave lower than 70% recovery. The low recoveries may have been due to instrument variability.

The uridine calibration curve for the combined injection method is shown in Figure B.4. This data was generated with four uridine solutions, two each at 0.00053 and 0.00106 mM, prepared in methanol. Methanol was added as described above.

A photohydrate calibration curve was needed, but a commercially available photohydrate standard was not available. Therefore, a photohydrate standard was created by exposing a 0.012 mM uridine solution to UV light and the resulting calibration curve is shown in Figure B.5. The solution was exposed to light from the medium-pressure UV lamp until a 71 percent uridine reduction was achieved. Serial dilutions (2, 4, and 8 fold) of the stock solution provided varied photohydrate concentrations and the photohydrate concentration was estimated by using equation B.2. The three ions corresponding to the caffeine standard, uridine, and the photohydrate at m/z 249, 267, and 285 were acquired simultaneously and integrated to provide corresponding peak areas. The concentration of the photohydrate in the diluted solutions was calculated (based on the dilution factor) and the concentration of the photohydrate was plotted versus the ratio of the area of the photohydrate divided by the area of the caffeine standard.

The alternating injection method involved injecting the caffeine standard (0.0026 mM caffeine in methanol) in an alternating fashion with the uridine samples (two samples of each uridine solution strength 0.00053 and 0.00106 mM). Sample injection order consisted of alternating uridine solution strength. As before, all three ions (m/z 249, 267, and 285) were acquired simultaneously and integrated to give the corresponding peak areas and the uridine calibration curve was generated (Figure B.6). Photohydrate standards were prepared similarly to what was described previously and were also introduced into the ES source in an alternating fashion with the standard. The photohydrate calibration curve that was produced (see Figure B.7) was determined in a manner similar to what was described for the combined injection method.

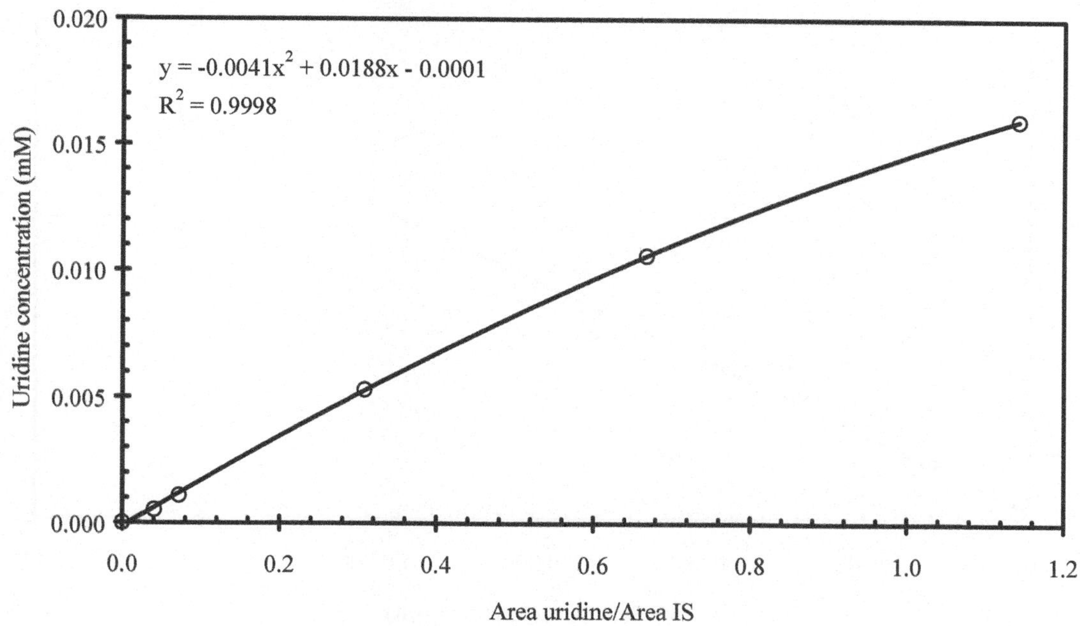

Figure B.4 Uridine calibration curve for the caffeine standard combined with samples

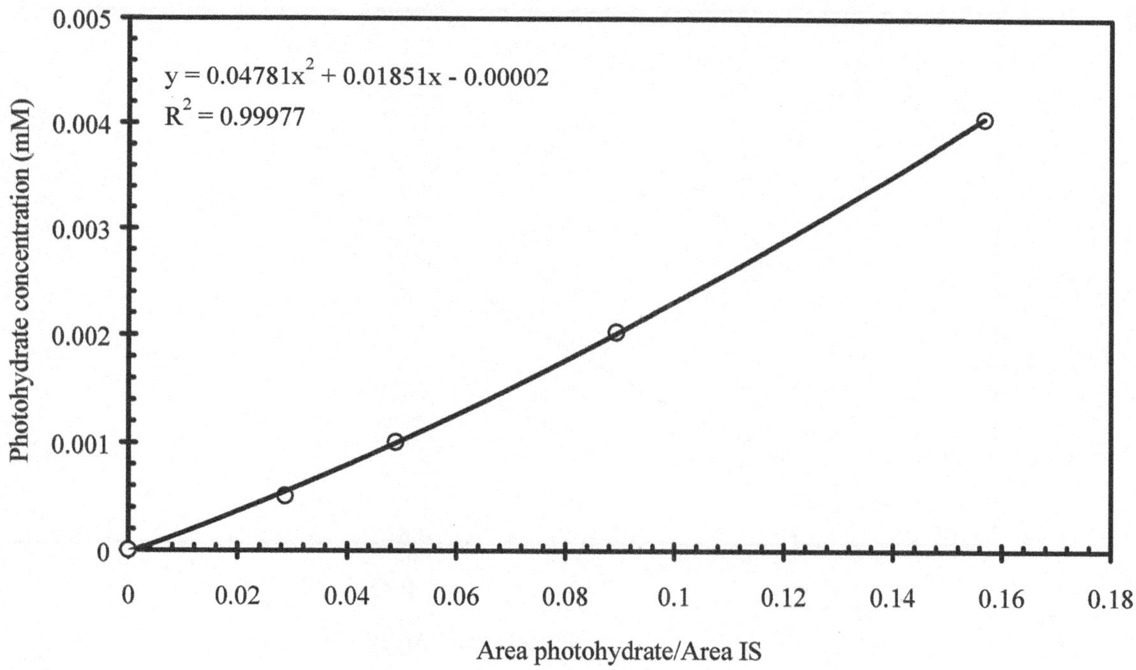

Figure B.5 Photohydrate calibration curve for the caffeine standard combined with samples

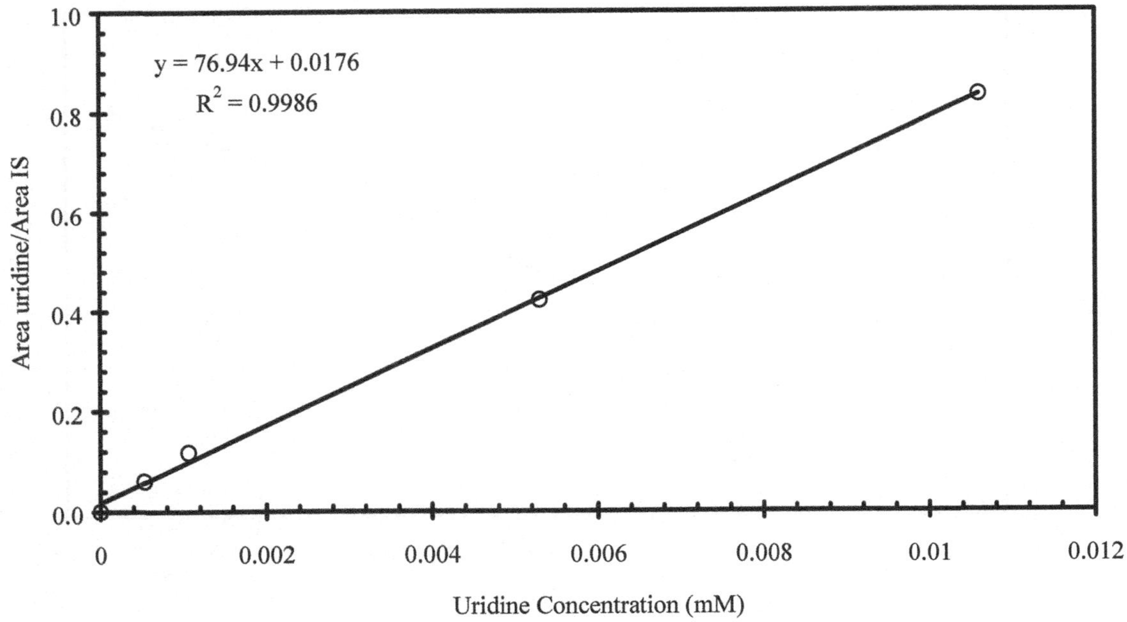

Figure B.6 Uridine calibration curve from alternating standard and sample injections

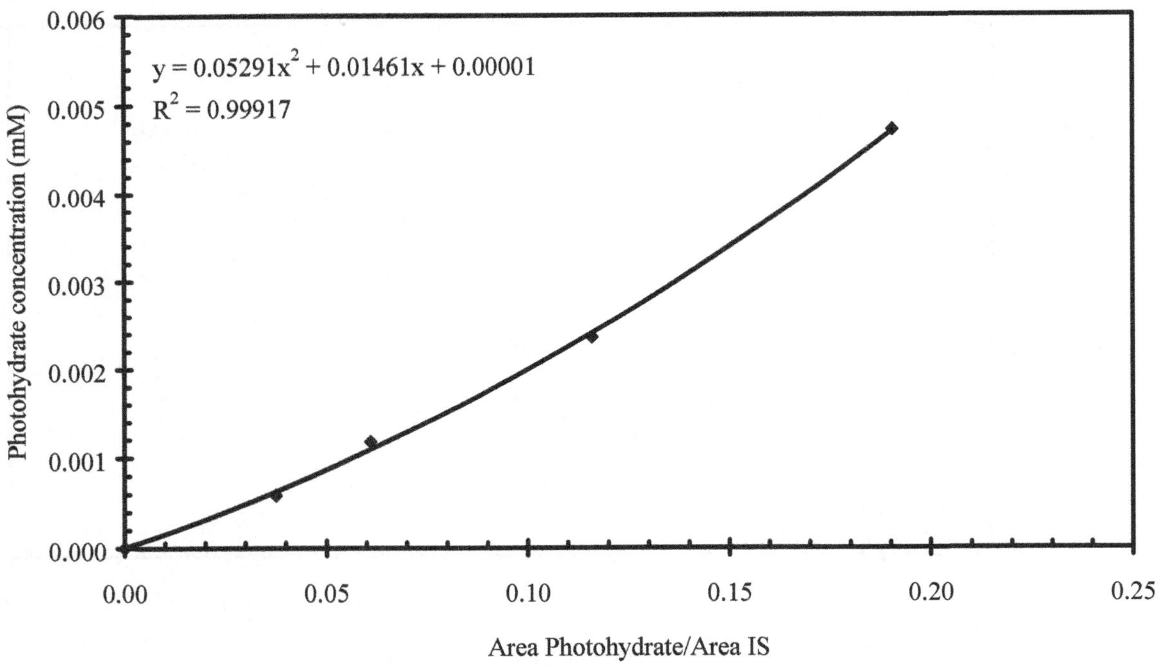

Figure B.7 Photohydrate calibration curve from alternating standard and sample injections

APPENDIX C
ACTINOMETER STABILITY TESTING METHODS

CHEMICALS

All materials used were reagent grade and used without further purification. Powdered uridine (Sigma Chemicals) was used in making uridine actinometer. Potassium iodide (KI) and potassium iodate (KIO_3) (Fisher Scientific Products) were used to prepare iodide and iodate ion (referred to as the KI actinometer in the following discussion).

PREPARATION OF ACTINOMETER SOLUTIONS

Uridine actinometer was prepared according to the method described by von Sonntag and Schuchmann (1992). Initial uridine concentration was 0.01mM. Solution was controlled at pH 7 by phosphate buffer. Iodide/iodate actinometer was prepared according to the method developed by Rahn (1997). Solutions of 0.6M KI and 0.1 M KIO_3 in 0.01M borate buffer (pH 9.2) were prepared as the actinometer. Sterilized, deionized water was used in the preparation of all solutions.

IRRADIATION

Uridine actinometer was irradiated under a LP-UV collimated beam system equipped with four 253.7nm mercury UV lamps (G15T8 15W Sankyo Denki, Japan). The target incident doses were 100, 500, 1000 mJ/cm^2. Unirradiated actinometer solution was used as the control sample for a dose of 0. The KI actinometer was irradiated under the same UV lamps for targeted incident doses of 10, 100, 1000 mJ/cm^2. Unirradiated solution was used as the control for a dose of 0.

STORAGE CONDITION TESTING

The four environmental conditions tested were as follows: (1) cold storage: duplicate aliquots of 15ml unirradiated and irradiated actinometer samples were stored in Quapak glass bottles at 5 °C (± 2 °C). (2) ambient dark storage: duplicate aliquots of 15ml unirradiated and irradiated actinometer samples were stored in amber glass bottles in the dark at 22 °C (± 2 °C). (3) ambient light storage: duplicate aliquots of 15ml unirradiated and irradiated actinometer samples were stored in clear glass bottles under fluorescent room light at 22 °C (± 2 °C). (4) warm storage: duplicate aliquots of 15ml unirradiated and irradiated actinometer samples were stored in amber glass bottles in the dark at 37 °C (± 1 °C). All the bottles were sterilized before using to prevent bacteria growth during storage.

SAMPLE ANALYSIS

For uridine actinometer, a sample of 1.3 ml was withdrawn from each bottle following 0, 4, 8, 12, 24, 72 (3 days), 120 (5 days), 168 (7 days) hours. All samples collected were analyzed immediately after withdrawal. For KI actinometer, samples were withdrawn at the same time periods as the uridine. However, for doses of 0 and 10 mJ/cm^2, a sample of 1.3 ml was withdrawn for direct analysis while for UV doses of 100 and 1000 mJ/cm^2, 0.25ml samples were withdrawn and diluted with deionized water 20 times and 100 times before analysis, respectively. The dilutions helped to bring the sample concentration to within detection limit of the spectrophotometer. Sample absorbances were measured using a Varian Cary 100 Bio-spectrophotometer immediately after withdrawal and, if necessary, dilution. All measurements were made relative to a baseline of deionized water.

APPENDIX D
SUPPLEMENTAL DATA

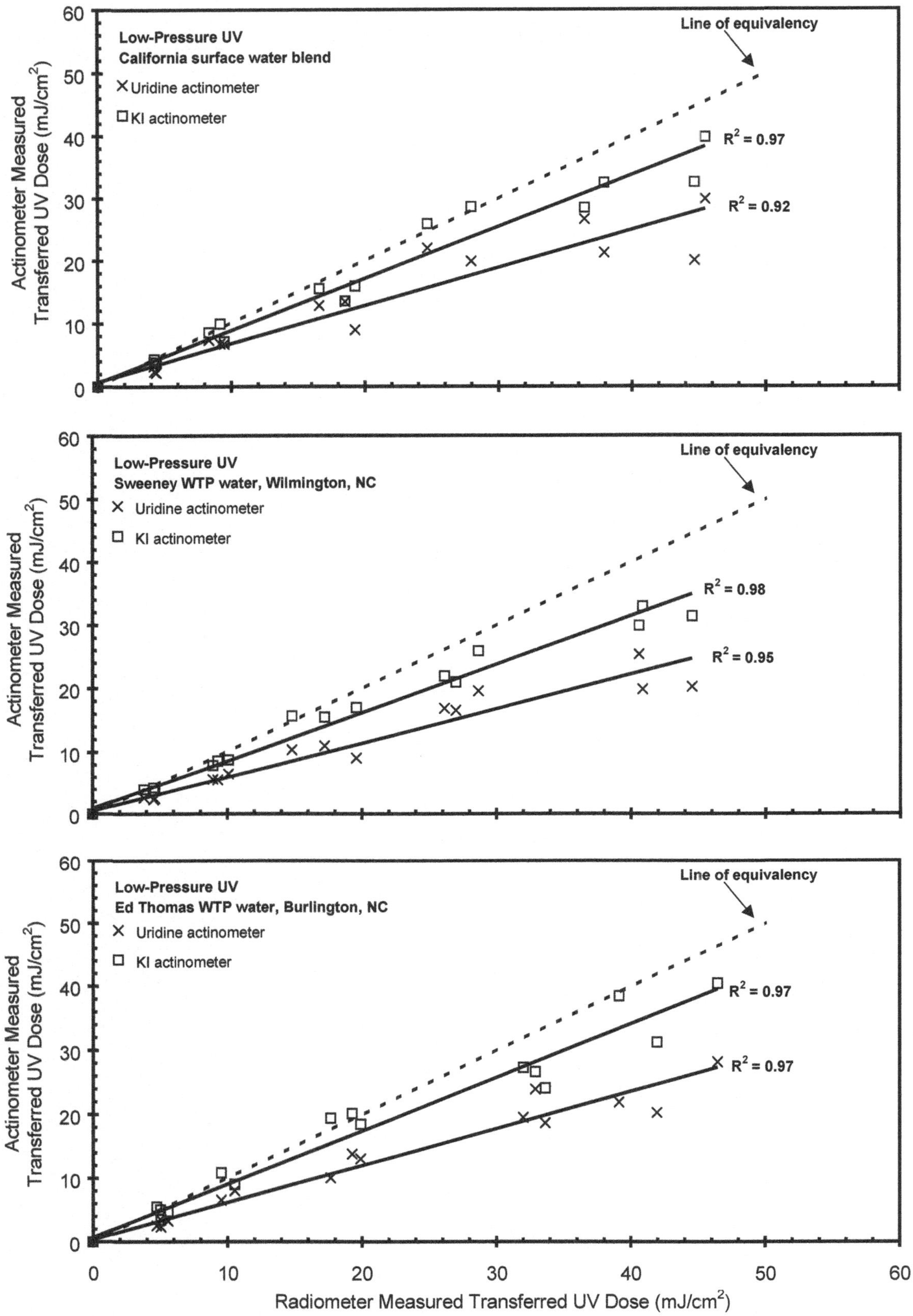

Low-pressure UV actinometer and radiometer comparison in natural waters

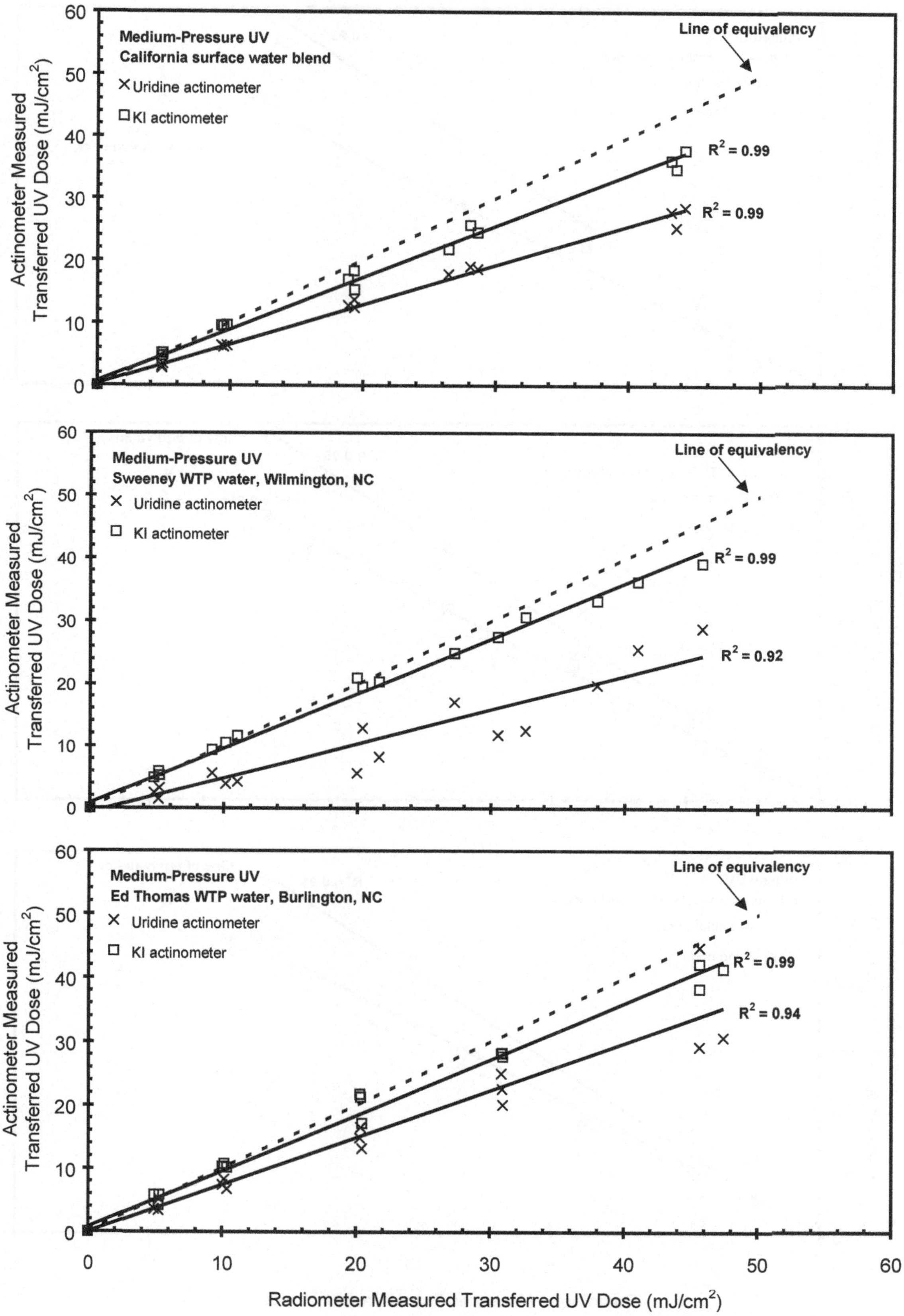

Medium-pressure UV actinometer and radiometer comparison in natural waters

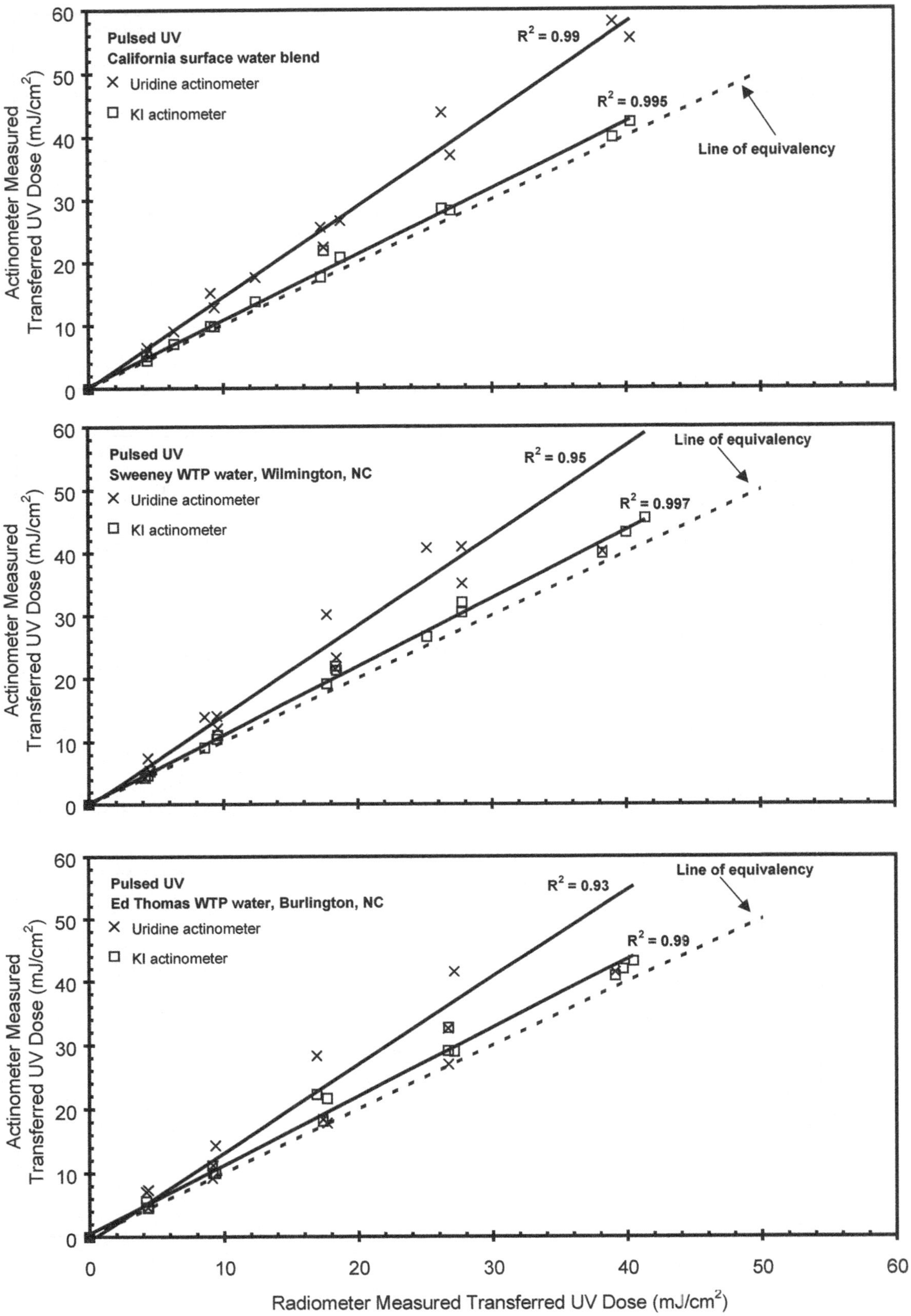

Pulsed UV actinometer and radiometer comparison in natural waters

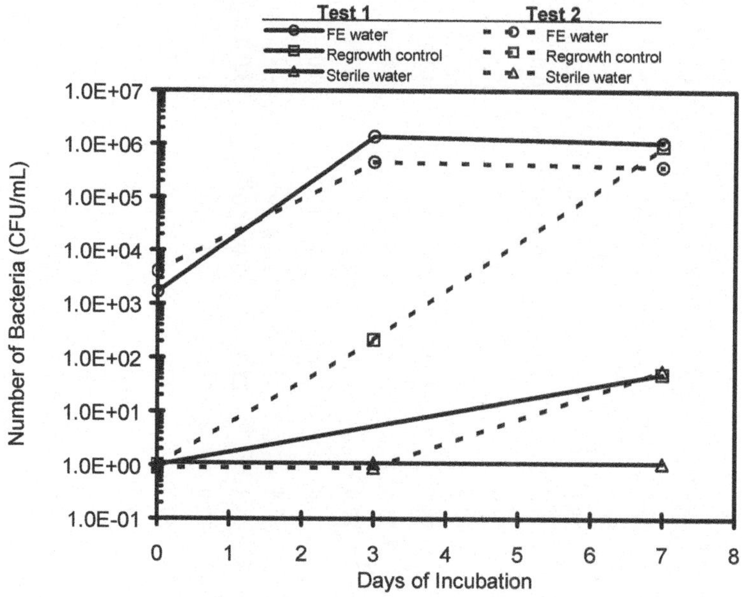

Control samples for light-exposure tests

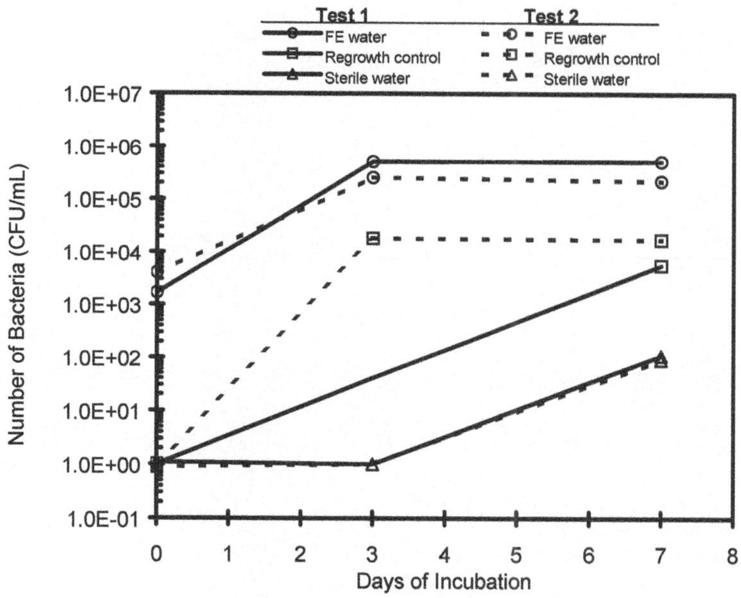

Control samples for dark-exposure tests

Positive and negative control samples for the light (uncovered reservoir simulated) and dark (distribution system simulated) UV repair/regrowth tests with heterotrophic bacteria

Raw data from biological dose measurement testing with *Bacillus subtilis* suspended in buffered water and exposed to UV light from the low-pressure lamp

Phase/ Test Type	UV Dose Radiometer (mJ/cm²)	UV Dose KI (mJ/cm²)	UV Dose Uridine (mJ/cm²)	Number of Bacteria (CFU/mL)	Calculated Inactivation Value (log)	Phase/ Test Type	UV Dose Radiometer (mJ/cm²)	UV Dose KI (mJ/cm²)	UV Dose Uridine (mJ/cm²)	Number of Bacteria (CFU/mL)	Calculated Inactivation Value (log)
1/1	0.0	-	-	2.2×10^6	0.0	1/5	0.0	-	-	2.3×10^6	0.0
	9.7	-	-	6.6×10^4	1.5		9.4	8.0	-	2.7×10^5	0.9
	38.5	-	-	9.0×10^1	4.4		19.0	15.8	-	9.0×10^2	3.4
	68.0	-	-	3.4×10^1	4.8		42.5	33.8	-	7.3×10^1	4.4
	98.8	-	-	1.5×10^1	5.2		66.2	52.6	-	3.3×10^1	4.8
	128.7	-	-	6.0×10^0	5.6		95.3	86.1	-	1.1×10^1	5.3
1/2	0.0	0.0	-	2.3×10^6	0.0	1/6	0.0	0.0	0.0	2.0×10^6	0.0
	9.0	6.2	-	9.5×10^4	1.3		10.0	9.6	5.3	8.8×10^4	1.4
	36.0	22.8	-	7.6×10^1	4.4		19.5	17.3	10.4	4.0×10^2	3.7
	63.2	38.2	-	3.3×10^1	4.8		43.5	38.2	23.2	7.0×10^1	4.5
	91.2	52.1	-	1.7×10^1	5.2		67.7	55.4	36.2	3.0×10^1	4.8
	121.5	62.5	-	7.0×10^0	5.5		93.5	77.9	51.8	$<1.0 \times 10^1$	>5.3
1/3	0.0	-	-	2.1×10^6	0.0	1/7	0.0	0.0	0.0	2.2×10^6	0.0
	9.6	7.8	-	1.2×10^5	1.2		10.5	8.4	13.2	1.5×10^4	2.2
	18.6	13.7	-	2.7×10^2	3.9		21.1	16.8	25.3	2.9×10^2	3.9
	42.6	29.5	-	3.7×10^1	4.8		47.7	37.9	51.5	1.1×10^2	4.3
	66.9	43.9	-	2.0×10^1	5.1		73.3	59.0	77.4	4.1×10^1	4.7
	96.2	57.9	-	6.0×10^0	5.6		99.1	84.0	106.9	$<1.1 \times 10^1$	>5.4
1/4	0.0	-	-	2.0×10^6	0.0	1/8	0.0	0.0	0.0	2.2×10^6	0.0
	9.7	6.4	-	8.9×10^4	1.4		9.8	5.8	9.0	9.2×10^4	1.4
	19.7	11.8	-	1.6×10^2	4.1		19.5	11.5	16.2	3.8×10^2	3.8
	48.8	28.9	-	2.8×10^1	4.9		44.2	25.8	33.8	1.3×10^2	4.2
	69.0	40.2	-	1.9×10^1	5.0		68.4	40.1	58.4	4.9×10^1	4.7
	97.8	47.8	-	1.6×10^1	5.1		96.5	57.3	82.1	1.9×10^1	5.1

Raw data from biological dose measurement with *Bacillus subtilis* suspended in buffered water and exposed to all three UV lamps

Phase/ Test/Lamp Type	UV Dose			Number of Bacteria (CFU/mL)	Calculated Inactivation Value (log)	Phase/ Test/Lamp Type	UV Dose			Number of Bacteria (CFU/mL)	Calculated Inactivation Value (log)
	Radiometer (mJ/cm^2)	KI (mJ/cm^2)	Uridine (mJ/cm^2)				Radiometer (mJ/cm^2)	KI (mJ/cm^2)	Uridine (mJ/cm^2)		
2/1/LP-UV	0.0	0.0	0.0	2.7×10^6	0.0	2/4/LP-UV	0.0	0.0	0.0	2.5×10^6	0.0
	9.5	8.5	5.0	1.1×10^5	1.4		21.0	16.8	6.8	6.2×10^2	3.5
	97.4	80.6	50.5	2.5×10^1	5.0		70.4	54.4	23.6	1.1×10^2	4.3
MP-UV	0.0	0.0	0.0	2.9×10^6	0.0		101.6	76.9	33.9	5.5×10^1	4.7
	10.3	9.9	4.3	1.7×10^3	3.2	MP-UV	0.0	0.0	0.0	2.3×10^6	0.0
	93.2	74.8	39.2	1.0×10^0	>6.3		22.1	20.2	10.9	1.7×10^2	4.2
P-UV	0.0	0.0	0.0	2.8×10^6	0.0		78.6	49.1	38.4	2.0×10^0	>6.1
	10.1	7.0	8.4	1.9×10^4	2.3		113.8	85.7	54.5	1.0×10^0	>6.9
	97.3	67.3	81.6	4.0×10^0	>6.0	P-UV	0.0	0.0	0.0	2.4×10^6	0.0
2/2/LP-UV	0.0	0.0	0.0	3.1×10^6	0.0		17.9	15.8	27.6	1.3×10^1	5.3
	18.7	17.4	15.7	1.2×10^3	3.4		63.1	55.8	97.3	$<1.0 \times 10^0$	>6.9
	42.4	36.2	35.2	2.4×10^2	4.2		89.9	79.5	138.7	$<1.0 \times 10^0$	>6.9
	68.2	57.5	54.7	1.4×10^2	4.4	2/5/LP-UV	0.0	0.0	0.0	3.3×10^6	0.0
MP-UV	0.0	0.0	0.0	2.6×10^6	0.0		19.9	19.4	12.9	1.1×10^3	3.5
	18.5	17.5	14.2	1.0×10^2	4.4		69.0	45.0	61.9	1.3×10^2	4.4
	42.5	36.4	32.5	6.0×10^0	>5.7	MP-UV	0.0	0.0	0.0	2.8×10^6	0.0
	64.0	55.9	49.9	6.0×10^0	>5.7		21.6	22.4	16.1	1.5×10^2	4.3
P-UV	0.0	0.0	0.0	3.1×10^6	0.0		75.2	69.1	56.5	3.0×10^0	>6.0
	19.3	14.1	24.4	4.0×10^2	3.9	P-UV	0.0	0.0	0.0	3.4×10^6	0.0
	43.5	31.9	55.1	5.4×10^1	4.8		21.6	36.2	50.8	6.8×10^1	4.6
	67.8	49.7	85.8	$<1.0 \times 10^1$	>5.4		76.3	104.2	179.3	1.0×10^0	>6.4
2/3/LP-UV	0.0	0.0	0.0	3.7×10^6	0.0	2/6/LP-UV	0.0	0.0	0.0	2.2×10^6	0.0
	9.5	9.0	5.6	2.1×10^5	1.1		10.0	10.2	5.4	2.2×10^5	1.2
	42.7	35.7	25.3	3.1×10^2	4.0		44.3	35.9	24.3	2.0×10^2	4.2
MP-UV	0.0	0.0	0.0	3.6×10^6	0.0		101.3	79.8	54.0	7.1×10^1	4.6
	18.5	17.3	9.6	1.6×10^2	4.3	MP-UV	0.0	0.0	0.0	2.5×10^6	0.0
	42.6	35.5	22.2	7.0×10^0	>5.7		10.6	10.1	7.8	3.0×10^3	2.9
P-UV	0.0	0.0	0.0	3.4×10^6	0.0		47.8	42.3	35.6	5.0×10^0	>5.7
	10.0	7.8	9.3	7.1×10^4	1.6		105.7	85.1	78.7	$<1.0 \times 10^0$	>6.9
	43.9	34.2	40.9	7.7×10^1	4.6	P-UV	0.0	0.0	0.0	2.6×10^6	0.0
							11.7	14.1	9.6	$<8.0 \times 10^3$	>2.5
							51.0	61.4	41.5	8.6×10^1	4.5
							113.1	136.1	92.0	$<1.0 \times 10^1$	>5.4

Raw data from biological dose measurement with *Bacillus subtilis* suspended in pretreated California surface water and exposed to all three UV lamps

Phase/ Test/Lamp Type	UV Dose			Number of Bacteria (CFU/mL)	Calculated Inactivation Value (log)	Phase/ Test/Lamp Type	UV Dose			Number of Bacteria (CFU/mL)	Calculated Inactivation Value (log)
	Radiometer (mJ/cm^2)	KI (mJ/cm^2)	Uridine (mJ/cm^2)				Radiometer (mJ/cm^2)	KI (mJ/cm^2)	Uridine (mJ/cm^2)		
3/1/LP-UV	0.0	0.0	0.0	2.7×10^6	0.0	3/4/MP-UV	0.0	0.0	0.0	2.6×10^6	0.0
	4.2	4.3	2.4	1.6×10^6	0.2		9.4	9.6	6.4	9.4×10^4	1.5
	37.9	32.5	21.2	1.3×10^2	4.2		28.1	25.7	19.0	3.3×10^2	4.0
MP-UV	0.0	0.0	0.0	2.3×10^6	0.0		44.3	37.8	28.5	3.9×10^1	4.9
	4.9	4.8	2.8	8.0×10^5	0.5	P-UV	0.0	0.0	0.0	2.8×10^6	0.0
	43.6	34.7	25.3	4.3×10^1	4.7		9.4	9.8	12.9	6.7×10^3	2.6
3/2/LP-UV	0.0	0.0	0.0	3.8×10^6	0.0		27.0	28.2	37.0	7.1×10^1	4.6
	9.5	7.1	6.7	5.3×10^5	0.8		40.4	42.3	55.5	3.5×10^1	4.9
	18.5	13.5	13.4	6.0×10^3	2.7	3/5/LP-UV	0.0	0.0	0.0	2.1×10^6	0.0
	36.5	28.5	26.6	2.4×10^2	4.1		8.3	8.5	7.3	3.5×10^5	0.8
MP-UV	0.0	0.0	0.0	2.7×10^6	0.0		24.7	25.9	22.0	5.9×10^2	3.6
	9.8	9.7	6.4	6.0×10^4	1.6	MP-UV	0.0	0.0	0.0	1.9×10^6	0.0
	18.9	17.0	12.8	2.9×10^2	3.9		9.5	6.2	9.5	3.9×10^4	1.7
	26.5	21.8	17.8	9.2×10^1	4.4		28.7	24.5	18.6	8.1×10^1	4.4
P-UV	0.0	0.0	0.0	3.1×10^6	0.0	P-UV	0.0	0.0	0.0	1.8×10^6	0.0
	6.4	7.1	9.1	3.4×10^5	0.9		9.1	9.9	15.1	8.3×10^3	2.3
	12.4	13.8	17.7	3.0×10^3	3.0		26.3	28.5	43.8	1.2×10^2	4.2
	18.8	20.8	26.7	2.1×10^2	4.2	3/6/LP-UV	0.0	0.0	0.0	2.4×10^6	0.0
3/3/LP-UV	0.0	0.0	0.0	2.6×10^6	0.0		4.4	3.8	2.2	9.0×10^5	0.4
	4.2	3.8	3.2	1.7×10^6	0.2		19.3	15.9	8.9	2.5×10^3	3.0
	16.6	15.5	12.8	2.4×10^3	3.1		44.7	32.5	20.0	2.3×10^2	4.0
MP-UV	0.0	0.0	0.0	2.7×10^6	0.0	MP-UV	0.0	0.0	0.0	2.3×10^6	0.0
	4.9	3.5	5.2	1.2×10^6	0.3		4.8	5.2	3.1	5.8×10^5	0.7
	19.3	18.3	13.7	2.1×10^2	4.1		19.4	15.3	12.4	4.5×10^2	3.8
P-UV	0.0	0.0	0.0	2.5×10^6	0.0		43.2	36.1	27.8	6.0×10^1	4.6
	4.3	5.4	5.6	8.3×10^5	0.5	P-UV	0.0	0.0	0.0	2.1×10^0	0.0
	17.5	21.9	22.4	2.9×10^2	3.9		4.4	4.4	6.5	4.0×10^5	0.8
3/4/LP-UV	0.0	0.0	0.0	3.1×10^6	0.0		17.3	17.7	25.6	3.8×10^2	3.8
	9.2	9.9	6.7	2.3×10^5	1.1		39.1	58.1	39.9	7.3×10^1	4.5
	28.0	28.6	19.9	1.2×10^2	4.4						
	45.5	39.8	29.8	8.1×10^1	4.6						

Raw data from biological dose measurement with *Bacillus subtilis* suspended in Sweeney water and exposed to all three UV lamps

Phase/ Test/Lamp Type	UV Dose Radiometer (mJ/cm^2)	UV Dose KI (mJ/cm^2)	UV Dose Uridine (mJ/cm^2)	Number of Bacteria (CFU/mL)	Calculated Inactivation Value (log)	Phase/ Test/Lamp Type	UV Dose Radiometer (mJ/cm^2)	UV Dose KI (mJ/cm^2)	UV Dose Uridine (mJ/cm^2)	Number of Bacteria (CFU/mL)	Calculated Inactivation Value (log)
4/1/LP-UV	0.0	0.0	0.0	3.0×10^6	0.0	4/4/LP-UV	0.0	0.0	0.0	2.8×10^6	0.0
	4.6	3.7	2.1	2.0×10^6	0.2		8.9	7.8	5.6	3.2×10^5	1.0
	40.9	32.9	18.9	5.2×10^2	3.7		26.1	21.9	16.8	5.7×10^2	3.7
MP-UV	0.0	0.0	0.0	3.2×10^6	0.0		40.6	25.3	29.9	3.9×10^2	3.9
	4.8	4.9	2.5	8.7×10^5	0.5	MP-UV	0.0	0.0	0.0	2.8×10^6	0.0
	38.0	33.2	19.8	1.6×10^2	4.3		9.1	9.4	5.7	4.2×10^4	1.9
P-UV	0.0	0.0	0.0	3.0×10^6	0.0		27.3	25.0	17.0	2.3×10^2	4.1
	4.2	4.3	4.4	7.1×10^5	0.7		41.0	36.3	25.6	8.2×10^1	4.5
	38.3	39.9	40.2	9.9×10^1	4.5	P-UV	0.0	0.0	0.0	2.7×10^6	0.0
4/2/LP-UV	0.0	0.0	0.0	2.8×10^6	0.0		8.6	9.1	13.9	1.5×10^4	2.3
	9.3	8.6	5.5	2.9×10^5	1.0		25.2	26.6	40.8	1.8×10^2	4.2
	17.2	15.5	11.0	5.6×10^3	2.7	4/5/LP-UV	0.0	0.0	0.0	3.4×10^6	0.0
	27.0	21.0	16.5	3.8×10^2	3.9		10.1	8.8	6.5	1.6×10^5	1.3
MP-UV	0.0	0.0	0.0	2.9×10^6	0.0		28.7	25.9	19.6	5.0×10^2	3.8
	11.0	11.6	4.2	4.3×10^3	2.8	MP-UV	0.0	0.0	0.0	2.8×10^6	0.2
	21.6	20.4	8.3	2.6×10^2	4.0		10.1	10.5	3.9	7.5×10^3	2.6
	32.6	30.7	12.5	8.4×10^1	4.5		30.5	27.6	11.8	9.3×10^1	4.5
P-UV	0.0	0.0	0.0	2.5×10^6	0.0	P-UV	0.0	0.0	0.0	3.3×10^6	0.0
	9.6	11.1	12.1	6.5×10^3	2.6		9.5	10.5	14.0	2.7×10^3	3.1
	18.4	21.3	23.3	2.8×10^2	4.0		27.8	30.4	40.9	1.3×10^2	4.4
	27.8	32.1	35.1	1.3×10^2	4.3		41.5	45.5	61.1	4.8×10^1	4.8
4/3/LP-UV	0.0	0.0	0.0	2.5×10^6	0.0	4/6/LP-UV	0.0	0.0	0.0	2.6×10^6	0.0
	3.8	4.0	2.6	1.2×10^6	0.4		4.5	4.2	2.3	9.9×10^5	0.4
	14.8	15.7	10.3	2.1×10^4	2.1		19.6	17.0	9.0	8.3×10^2	3.4
MP-UV	0.0	0.0	0.0	2.4×10^6	0.1		44.5	31.4	20.2	2.7×10^2	4.0
	5.1	6.0	1.4	6.6×10^5	0.7	MP-UV	0.0	0.0	0.0	2.0×10^6	0.0
	20.0	21.0	5.7	2.8×10^2	4.1		5.2	5.3	3.2	4.0×10^5	0.8
P-UV	0.0	0.0	0.0	3.1×10^6	0.0		20.4	12.9	19.6	4.2×10^2	3.8
	4.5	5.4	5.3	3.9×10^5	0.9		45.8	28.9	39.2	3.1×10^1	4.8
	18.4	21.9	21.6	3.2×10^2	3.9	P-UV	0.0	0.0	0.0	2.7×10^6	0.0
							4.4	4.7	7.4	3.1×10^5	0.9
							17.7	19.1	30.1	3.5×10^2	3.9
							40.0	43.2	68.1	5.3×10^1	4.7

Raw data from biological dose measurement with *Bacillus subtilis* suspended in Duke water and exposed to all three UV lamps

Phase/ Test/Lamp Type	UV Dose Radiometer (mJ/cm^2)	KI (mJ/cm^2)	Uridine (mJ/cm^2)	Number of Bacteria (CFU/mL)	Calculated Inactivation Value (log)	Phase/ Test/Lamp Type	UV Dose Radiometer (mJ/cm^2)	KI (mJ/cm^2)	Uridine (mJ/cm^2)	Number of Bacteria (CFU/mL)	Calculated Inactivation Value (log)
5/1/LP-UV	0.0	0.0	0.0	2.0×10^6	0.1	5/4/LP-UV	0.0	0.0	0.0	3.0×10^6	0.0
	5.0	3.5	2.3	5.7×10^5	0.7		5.6	4.7	3.3	7.3×10^5	0.6
	41.9	31.3	20.3	2.3×10^2	4.1		33.6	24.1	18.7	3.8×10^2	3.9
MP-UV	0.0	0.0	0.0	2.3×10^6	0.0		46.4	40.4	28.2	2.5×10^2	4.1
	5.3	5.8	5.1	5.0×10^5	0.7	MP-UV	0.0	0.0	0.0	2.6×10^6	0.0
	45.7	42.2	44.7	2.0×10^1	5.1		10.4	10.2	6.8	3.5×10^3	2.9
P-UV	0.0	0.0	0.0	3.0×10^6	0.0		31.0	27.7	20.1	4.7×10^1	4.7
	4.4	4.6	7.3	3.9×10^5	0.9		47.4	41.4	30.7	8.0×10^0	5.5
	39.7	42.0	66.9	5.3×10^1	4.7	P-UV	0.0	0.0	0.0	2.5×10^6	0.0
5/2/LP-UV	0.0	0.0	0.0	2.1×10^6	0.0		9.4	10.0	14.3	3.6×10^3	2.8
	9.5	10.8	6.6	1.4×10^5	1.2		27.1	29.0	41.5	1.1×10^2	4.3
	19.9	18.5	13.1	9.7×10^2	3.3		40.5	43.2	61.9	3.9×10^1	4.8
	32.0	27.3	19.6	3.6×10^2	3.8	5/5/LP-UV	0.0	0.0	0.0	2.3×10^6	0.0
MP-UV	0.0	0.0	0.0	2.3×10^6	0.0		10.5	9.0	8.0	6.0×10^4	1.7
	10.2	10.8	8.3	8.8×10^3	2.5		32.9	26.7	24.0	3.4×10^2	3.9
	20.4	21.3	16.5	2.6×10^2	4.0	MP-UV	0.0	0.0	0.0	2.8×10^6	0.0
	30.9	25.0	28.2	5.7×10^1	4.6		10.0	10.3	7.4	4.7×10^3	2.8
P-UV	0.0	0.0	0.0	2.2×10^6	0.0		30.9	28.3	22.6	6.5×10^1	4.6
	9.1	11.2	9.2	2.8×10^3	2.9	P-UV	0.0	0.0	0.0	2.8×10^6	0.0
	17.7	21.6	17.8	2.4×10^2	4.0		9.2	10.0	11.2	4.2×10^5	0.8
	26.7	32.7	26.9	8.5×10^1	4.4		26.7	29.1	32.6	1.3×10^2	4.4
5/3/LP-UV	0.0	0.0	0.0	2.1×10^6	0.0	5/6/LP-UV	0.0	0.0	0.0	3.1×10^6	0.0
	5.0	5.0	3.5	8.2×10^5	0.4		4.7	5.5	2.5	1.0×10^6	0.5
	19.3	20.2	13.8	9.0×10^2	3.4		17.7	19.4	10.1	1.2×10^3	3.4
MP-UV	0.0	0.0	0.0	2.3×10^6	0.0		39.1	38.5	21.9	2.8×10^2	4.1
	4.9	5.9	3.6	4.7×10^5	0.8	MP-UV	0.0	0.0	0.0	2.8×10^6	0.0
	20.3	21.8	14.9	2.6×10^2	4.0		5.2	4.2	3.4	6.0×10^5	0.7
P-UV	0.0	0.0	0.0	2.6×10^6	0.0		20.4	17.2	13.2	2.8×10^2	4.0
	4.2	5.5	7.1	5.3×10^5	0.7		45.6	38.3	29.3	2.4×10^1	5.1
	16.9	22.2	28.3	3.3×10^2	3.9	P-UV	0.0	0.0	0.0	2.4×10^6	0.0
							4.3	4.5	4.6	6.2×10^5	0.6
							17.4	18.1	18.4	3.2×10^2	3.9
							39.1	40.8	41.4	6.0×10^1	4.6

Raw data from disinfection testing with coliphage MS-2

Test No. (no.)	Lamp Type	Radiometric UV Dose (mJ/cm^2)	Number of Coliphage (PFU/mL)	Calculated Inactivation Value (log)	Test No. (no.)	Lamp Type	Radiometric UV Dose (mJ/cm^2)	Number of Coliphage (PFU/mL)	Calculated Inactivation Value (log)
1a.	LP-UV	0.0	7.4×10^5	0.0	3c.	P-UV	0.0	3.0×10^5	–0.1
		10.5	1.2×10^5	0.8			8.9	3.0×10^4	0.9
		30.3	6.6×10^3	2.1			39.1	1.3×10^2	3.3
1b.	MP-UV	0.0	7.1×10^5	0.0	4a.	LP-UV	0.0	2.0×10^6	0.0
		10.8	1.1×10^5	0.8			19.2	6.0×10^4	1.6
		32.0	4.3×10^3	2.2			30.2	1.2×10^4	2.3
1c.	P-UV	0.0	7.4×10^5	0.0			74.9	8.1×10^1	4.4
		8.6	5.1×10^4	1.1	4b.	MP-UV	0.0	2.0×10^6	0.0
		25.1	5.3×10^2	3.1			21.1	3.6×10^4	1.8
2a.	LP-UV	0.0	6.0×10^5	0.0			31.8	9.0×10^3	2.4
		22.9	1.3×10^4	1.6			75.0	2.1×10^1	5.0
		50.8	3.6×10^2	3.2	5a.	LP-UV	0.0	2.1×10^6	0.0
		77.7	2.4×10^1	4.4			19.4	3.6×10^4	1.8
2b.	MP-UV	0.0	5.5×10^5	0.0			67.1	2.1×10^2	4.0
		21.6	9.3×10^3	1.8	5b.	MP-UV	0.0	2.2×10^6	0.0
		48.4	4.3×10^2	3.2			21.5	3.1×10^4	1.9
		75.7	3.0×10^0	5.3			68.1	2.2×10^1	5.0
2c.	P-UV	0.0	4.8×10^5	0.0	6a.	LP-UV	0.0	1.9×10^6	0.0
		17.5	6.8×10^3	1.9			10.5	2.8×10^5	0.9
		39.5	1.7×10^2	3.5			30.7	1.9×10^4	2.0
		61.6	5.0×10^0	5.0			45.1	2.6×10^3	2.9
3a.	LP-UV	0.0	2.8×10^5	0.0	6b.	MP-UV	0.0	2.1×10^6	0.0
		9.5	5.2×10^4	0.8			10.8	2.6×10^5	0.9
		47.3	4.8×10^2	2.8			32.4	5.6×10^3	2.6
3b.	MP-UV	0.0	3.3×10^5	0.0			47.4	7.3×10^2	3.5
		11.0	3.8×10^4	0.9					
		48.6	1.8×10^1	3.2					

Raw data from disinfection testing with heterotrophic bacteria

Test No. (no.)	Lamp Type	Radiometric UV Dose (mJ/cm^2)	Number of Coliphage (PFU/mL)	Calculated Inactivation Value (log)	Test No. (no.)	Lamp Type	Radiometric UV Dose (mJ/cm^2)	Number of Coliphage (PFU/mL)	Calculated Inactivation Value (log)
1a.	LP-UV	0.0	1.3×104	0.0	5b.	MP-UV	0.0	4.0×103	0.0
		5.1	2.1×10^{3}	0.8			11.3	1.8×10^{1}	2.4
		56.8	1.0×10^{0} *	>4.0			38.7	1.0×10^{0} *	>3.6
1b.	MP-UV	0.0	1.6×10^{4}	0.0	6a.	LP-UV	0.0	5.7×10^{3}	0.1
		5.6	1.2×10^{3}	1.1			5.5	5.6×10^{2}	1.0
		60.0	$<1.0 \times 10^{0}$ *	>4.7			16.5	2.0×10^{0} *	>3.5
2a.	LP-UV	0.0	2.7×10^{3}	0.0			60.2	1.0×10^{0} *	>3.8
		9.8	1.4×10^{1}	2.3	6b.	MP-UV	0.0	5.7×10^{3}	0.0
		14.8	4.0×10^{0}	2.8			5.5	4.9×10^{2}	1.1
		33.9	1.0×10^{0} *	>3.3			16.3	4.0×10^{0}	3.2
2b.	MP-UV	0.0	2.4×10^{3}	0.0			59.4	$<1.0 \times 10^{0}$ *	>4.2
		10.6	1.2×10^{1}	2.3	7a.	P-UV	0.0	1.6×10^{4}	0.0
		15.8	5.0×10^{0}	2.7			5.0	7.2×10^{2}	1.4
		37.7	1.0×10^{0} *	>3.3			9.8	3.0×10^{0}	3.7
3a.	LP-UV	0.0	1.1×10^{4}	0.0			14.8	2.0×10^{0} *	>3.8
		5.7	1.4×10^{3}	0.9			34.7	2.0×10^{0} *	>3.9
		16.3	6.0×10^{0}	3.3			54.2	1.0×10^{0} *	>4.2
3b.	MP-UV	0.0	1.2×10^{4}	0.0	8a.	P-UV	0.0	2.7×10^{3}	0.0
		5.4	4.1×10^{2}	1.5			4.8	7.3×10^{1}	1.6
		16.2	3.0×10^{0}	3.6			9.5	3.0×10^{0}	3.0
4a.	LP-UV	0.0	7.3×10^{3}	0.0			14.5	2.0×10^{0} *	>3.2
		10.9	1.9×10^{1}	2.6			34.1	$<1.0 \times 10^{0}$ *	>3.6
		37.6	$<1.0 \times 10^{0}$ *	>4.3			53.8	$<1.0 \times 10^{0}$ *	>3.9
		57.8	$<1.0 \times 10^{0}$ *	>4.0	9a.	P-UV	0.0	7.4×10^{3}	0.0
4b.	MP-UV	0.0	7.2×10^{3}	0.0			4.8	1.8×10^{2}	1.6
		10.7	9.0×10^{0}	2.9			9.5	3.0×10^{0}	3.4
		37.9	1.0×10^{1}	2.8			14.7	2.0×10^{0}	3.6
		48.6	$<1.0 \times 10^{0}$ *	>4.0			34.5	1.0×10^{0} *	>3.9
5a.	LP-UV	0.0	6.2×10^{3}	0.0			54.2	$<1.0 \times 10^{0}$ *	>4.1
		11.3	2.3×10^{0}	3.4					
		39.4	1.0×10^{0} *	>3.7					

The number of bacteria/mL in the stock solutions, typically less than 5 percent higher than the 0-dose sample, is not shown.

* Samples with ≤ 2 CFU/mL were interpreted as below the detectable level

Raw data from disinfection testing with *Cryptosporidium parvum*

Test No. (no.)	Lamp Type	Radiometric UV Dose (mJ/cm^2)	Number of Oocysts (no.)	Calculated Log Inactivation (avg. ± st. dev.)
1	+ Control	0.0	1.6×10^4	0.0
	LP-UV	0.9	2.8×10^4	0.3 ± 0.0
		9.4	1.0×10^2	2.2 ± 0.3
	MP-UV	1.1	5.3×10^3	0.5 ± 0.0
		14.1	8.9×10^1	2.2 ± 0.3
	P-UV	0.2	2.4×10^4	0.0
		1.9	1.3×10^4	0.1 ± 0.3
2	+ Control	0.0	8.8×10^3	0.0
	LP-UV	2.6	4.2×10^3	0.3 ± 0.0
		4.4	3.5×10^2	1.4 ± 0.1
	MP-UV	3.4	8.1×10^2	1.0 ± 0.0
		5.5	4.0×10^1	2.3 ± 0.1
	P-UV	3.3	4.8×10^2	1.3 ± 0.0
		5.2	1.6×10^1	2.7 ± 0.3
3	+ Control	0.0	8.8×10^3	0.0
	LP-UV	0.8	6.8×10^3	0.1 ± 0.0
		6.7	1.6×10^2	1.7 ± 0.1
	MP-UV	1.1	2.4×10^3	0.6 ± 0.0
		8.9	1.9×10^1	2.7 ± 0.1
	P-UV	1.1	1.5×10^3	0.8 ± 0.0
		7.8	3.2×10^1	2.4 ± 0.3
4	+ Control	0.0	4.4×10^3	0.0
	LP-UV	2.6	1.6×10^3	0.4 ± 0.0
		4.4	1.7×10^2	1.4 ± 0.1
	MP-UV	3.5	2.3×10^2	1.3 ± 0.0
		5.6	3.2×10^1	2.1 ± 0.1
	P-UV	3.1	3.6×10^2	1.1 ± 0.0
		5.0	3.1×10^1	2.2 ± 0.3
5	+ Control	0.0	6.6×10^3	0.0
	LP-UV	8.2	6.3×10^1	2.0 ± 0.0
		11.8	3.6×10^1	2.3 ± 0.1
	MP-UV	8.7	4.0×10^1	2.2 ± 0.0
		12.2	4.9×10^1	2.1 ± 0.1
	P-UV	7.2	4.1×10^1	2.2 ± 0.3
		10.0	9.3×10^1	1.9 ± 0.3
6	+ Control	0.0	7.3×10^3	0.0
	LP-UV	3.1	3.4×10^3	0.3 ± 0.0
		7.8	5.4×10^1	2.1 ± 0.1
	MP-UV	3.3	8.9×10^2	0.9 ± 0.0
		8.6	3.6×10^1	2.3 ± 0.1
	P-UV	2.9	2.3×10^3	0.5 ± 0.0
		7.5	6.8×10^1	2.0 ± 0.3
7	+ Control	0.0	8.2×10^3	0.0
	LP-UV	1.2	3.1×10^3	0.4 ± 0.0
		5.7	9.8×10^1	1.9 ± 0.1
	MP-UV	1.8	3.3×10^3	0.4 ± 0.0
		8.2	9.0×10^1	2.0 ± 0.1
8	+ Control	0.0	6.3×10^3	0.0
	LP-UV	2.6	2.4×10^3	0.4 ± 0.0
		9.6	$<3.5 \times 10^1$	>2.3
	MP-UV	3.3	2.0×10^3	0.5 ± 0.0
		11.9	4.0×10^2	1.2 ± 0.1
9	+ Control	0.0	8.1×10^3	0.0
	LP-UV	0.5	$>8.8 \times 10^3$	0.0
		23.5	$<6.6 \times 10^1$	>2.1
	MP-UV	0.9	$>7.6 \times 10^3$	0.0
		26.8	$<6.6 \times 10^1$	>2.1

REFERENCES

Adams, M.H. 1959. *Bacteriophages*. New York: Interscience Publishers, Inc.

Arrowood M.J., L.T. Xie, K. Rieger, and J. Dunn. 1996. Disinfection of *Cryptosporidium parvum* Oocysts by Pulsed Light Treatment Evaluated in an in vitro Cultivation Model *J. Eukar. Microbiol.*, 43(5):S88

APHA, AWWA, and WEF (American Public Health Association, American Water Works Association, and Water Environment Federation). 1998. *Standard Methods for the Examination of Water and Wastewater*. 20th ed. Washington, D.C.: APHA.

Baron, J. 1997. Repair of Wastewater Microorganisms After Ultraviolet Disinfection Under Seminatural Conditions. *Wat. Environ. Res.*, 69(5):992-998.

Battigelli, D.A., M.D. Sobsey, and D.C. Lobe. 1993. The Inactivation of Hepatitis A Virus and Other Model Viruses by UV Irradiation. *Wat. Sci. Tech.*, 27(3-4):339-342.

Blaney, R., T. Al-Nakib, and R.J.H. Davies. 1993. A Flourogenic Actinometer for Nucleic Acid Photochemistry Utilizing 1-Deazapurine N(3)-Oxides. *Photochem. Photobiol.*, 57:380.

Bolton, J.R. 2000a. Calculation of Ultraviolet Fluence Rate Distributions in an Annular Reactor: Significance of Refraction and Reflection. *Wat. Res.*, 34(13):3315-3324.

Bolton, J.R. 2000b. Terms and Definitions in Ultraviolet Disinfection. In P*roc. of Disinfection 2000: Disinfection of Wastes in New Millennium, March 15-18, 2000, New Orleans, LA.* Alexandria, VA: Water Environment Federation.

Braunstein J.L., F.J.Loge, G. Tchobanoglous, and J.L. Darby. 1996. Ultraviolet Disinfection of Filtered Activated Sludge Effluent for Reuse Applications. *Wat. Environ. Res.*, 68(2):152-161.

Bukhari, Z., T.M. Hargy, J.R. Bolton, B. Dussert, and J.L. Clancy. 1999. Medium-Pressure UV Light for Oocyst Inactivation. *J. AWWA,* 91(3):86-94.

Cairns, W.L. 1996. Comparison of UV Disinfection Technologies Using Low Intensity Monochromatic and High Intensity Polychromatic UV Lamps. *In Proc. Disinfecting Wastewater for Discharge and Reuse, March 17-20, 1996, Portland, OR.* Alexandria, VA: Water Environment Federation.

Campbell, A.T., L.J. Robertson, and H.V. Smith. 1995. Inactivation of Oocysts of *Cryptosporidium parvum* by Ultraviolet Irradiation, *Wat. Res.*, 29:2583.

Carlson, D.A., R.W. Seabloom, F.B. DeWalle, T.F. Wetzler, J. Engeset, R. Butler, S. Wangsuphachart, and S. Wang. 1985. *Ultraviolet Disinfection of Water for Small Water Supplies.* EPA/600/S2-85/092, Cincinnati, OH: USEPA.

Chang, J.C.H., S.F. Osoff, D.C. Lobe, M.H. Dorfman, C.M. Dumais, R.G. Qualls and J.D. Johnson. 1985. UV Inactivation of Pathogenic and Indicator Microorganisms. *Appl. Environ. Microbiol.*, 49(6):1361-1365.

Chiu K.P., D.A. Lyn, P. Savoye, and E.R. Blatchley III. 1999. Integrated UV Disinfection Model Based on Particle Tracking. *ASCE: J. Environ. Eng.* 125(1):7-16.

Chrtek, S. and W. Popp. 1991. UV Disinfection of Secondary Effluents from Sewage Treatment Plants. *Wat. Sci. Tech.*, 24:343-346.

Clancy, J.L., T.M. Hargy, M.M. Marshall, and J.E. Dyksen. 1998a. *Innovative Electrotechnologies for Inactivation of Cryptosporidium parvum.* Denver, Colo.: AwwaRF and AWWA.

Clancy, J., T. Hargy, J. Dyksen. 1998b. UV Light Inactivation of *Cryptosporidium* Oocysts. *Jour. AWWA*, 90(9):92.

Clancy, J.L., Z. Bukhari, T.M. Hargy, J.R. Bolton, B.W. Dussert, and M.M. Marshall. 2000. Using UV to Inactivate *Cryptosporidium*. *J. AWWA*, 92(9):97–104.

Craik, S. A., D. Weldon, G.R. Finch, J.R. Bolton, and M. Belosevic. 2001. Inactivation of *Cryptosporidium parvum* Oocysts Using Medium- and Low-Pressure Ultraviolet Radiation. *Wat. Res.*, 35(6):1387-1398.

Darby, J.L., K.E. Snider, and G. Tchobanoglous. 1993. Ultraviolet Disinfection for Wastewater Reclamation and Reuse Subject to Restrictive Standards. *Wat. Environ. Res.*, 65(2)169.

Dunn J., T. Ott, and W. Clark. 1995. Pulsed-Light Treatment of Food and Packaging. *Food Technol.,* 49(9):95-98.

Finch, G.R., K.E. Black, L. Gyurek, and M. Belosevic. 1994. *Ozone Disinfection of* Giardia *and* Cryptosporidium. Denver, Colo.: AwwaRF and AWWA.

Gahr, A., L. Weil, and R. Neibner. 1995. Polychromatic Actinometry with Filter Solutions. *Wat. Res.,* 29:2125.

Gates, F.L. 1929. A Study of the Bactericidal Action of Ultra Violet Light, I. The Reaction to Monochromatic Radiation. *J. Gen. Physiol.*, 13:231.

Hargy, T.M. 1999. The Use of Ultraviolet Light for Inactivation of *Cryptosporidium* in Water. In *Proc. Water Quality Association Annual Conference*, Fort Worth, TX.: WQA.

Hargy, T.M. 2000. UV Equipment Proven Against *Cryptosporidium. Wat. Condition. Purific.,* March:36-38.

Harm, W. 1980. *Biological Effects of Ultraviolet Radiation.* New York: Cambridge University Press.

Harris, G.D., V.D. Adams, D.L. Sorensen and M.S. Curtis. 1987. Ultraviolet Inactivation of Selected Bacteria and Virus With Photoreactivation of the Bacteria. *Wat. Res.*, 21(6):687-692.

Hoyer, O. and K. Nick. 1999. *Working Procedure for Determining Microbiocidal Irradiance With Uridine Actinometry - Calibration of UV Sensors.* Wahnbachtelsperrenverband Siegburg, Germany. In German.

Huffman D.E., T.R. Slifko, K. Salisbury, J.B. Rose. 2000. Inactivation of Bacteria, Virus and Cryptosporidium by a Point-of-Use Device Using Pulsed Broad Spectrum White Light. *Wat. Res.,* 34(9):2491-2498.

Jagger, J.H. 1958. Photoreactivation. *Bact. Rev*., 22:99-142.

Jagger, J.H. 1967. *Introduction to Research in UV Photobiology.* Englewood Cliffs, N.J.: Prentice-Hall, Inc.

Karanis, P., W.A. Maier, H.M. Seitz, and D. Schoenen. 1992. UV Sensitivity of Protozoan Parasites. *J. Water SRT- Aqua*, 41:95-100.

Kuhn, H. J., S.E. Braslavsky, and R. Schmidt. 1989. Chemical Actinometry. *Pure Appl. Chem.,* 61:187-210.

Kwan, A., A. Mohammed, J. Archer, F. Soroushian, and G. Tchobanoglous. 1996. Factors for Selection of a High Intensity UV Disinfection System for a Large Scale Application. *In Proc. Disinfecting Wastewater for Discharge and Reuse, March 17-20, 1996, Portland, OR.* Alexandria, VA: Water Environment Federation.

Lin, L.S, C.T. Johnson, and E.R. Blatchley. 1997 Numerical Modeling of Inorganic Fouling at Quartz: Water Interfaces in Ultraviolet Photoreactors. *In Proc. CSCE/ASCE Environmental Engineering Conference, July 22-26, 1997, Edmonton, Alberta, Canada*: ASCE/CSCE.

Linden, K.G. and J.L., Darby. 1997. Estimating Effective Germicidal Dose From Medium Pressure UV Lamps. *ASCE: J. Environ. Eng.*, 123(11):1142-1149.

Linden, K.G. and J.L. Darby. 1998. UV Disinfection of Marginal Effluents: Determining UV Absorbance and Subsequent Estimation of UV Intensity. *Wat. Environ. Res.*, 70(2):214-223.

Linden, K.G., G.A. Shin, and M.D. Sobsey. 2000. Comparison of Monochromatic and Polychromatic UV Light for Disinfection Efficacy. In *Proc. Water Quality Technology Conference, November 4-8, 2000, Salt Lake City, UT.* Denver, Colo.: AWWA.

Linden, K.G., G.A. Shin, and M.D. Sobsey. 2001. Relative Efficacy of UV Wavelengths for the Inactivation of *Cryptosporidium parvum. Wat. Sci. Technol.*, 43(12):171.

Linden, K.G., G.A. Shin, G. Faubert, W. Cairns, M.D. Sobsey. 2002. UV Disinfection of *Giardia lamblia* in Water. *Environ. Sci. Technol.*, 36(11):2519.

Lindenauer, K.G. and J.L. Darby. 1994. Ultraviolet Disinfection of Wastewater: Effect of Dose on Subsequent Photoreactivation. *Wat. Res.*, 28(4):805-817.

Lyn D.A., K. Chiu, and E.R. Blatchley III. 1999. Numerical Modeling of Flow and Disinfection in UV Disinfection Channels. *ASCE: J Environ. Eng.*, 125(1):17-26.

Maier, A., D. Tougianidou, A. Wiedenmann, K. Botzenhart. 1995. Detection of Poliovirus by Cell Culture and by PCR After UV Disinfection. *Wat. Sci. Tech.*, 31(5-6):141-145.

Malley, J.P. 2000. Reduction of Microbial and Chemical Pollution of Coastal Waters Using Pulsed-UV Disinfection. University of New Hampshire, Durham, NH., Interim Report, NOAA.

Malley, J.P. Jr., J.R. Shaw, and J.D. Ropp. 1995. *Evaluation of the By-Products Produced by Treatment of Groundwaters with Ultraviolet Irradiation.* Denver, Colo.: AwwaRF and AWWA.

Meng, Q.S. and C.P. Gerba. 1996. Comparative Inactivation of Enteric Adenovirus, Poliovirus and Coliphages by Ultraviolet Irradiation. *Wat. Res.*, 30(11):2665-2668.

Mofidi, A.A., H. Baribeau, J.F. Green, R. De Leon, and R.L. Wolfe. 1998. Disinfection Using-High Intensity Pulsed-Ultraviolet Irradiation. In *Proc. Water Quality Technology Conference, November 1-5, 1998, San Diego, Calif.,* Denver, Colo.: AWWA.

Mofidi, A.A. H. Baribeau, P.A. Rochelle, R. De Leon, B.M. Coffey; and J.F. Green. 2001. Disinfection of *Cryptosporidium parvum* With Polychromatic UV Light. *Jour. AWWA*, 93(6)95.

Mofidi, A.A, E.A. Meyer, P.M. Wallis, C.I. Chou, B.P. Meyer, S. Ramalingam, and B.M. Coffey. 2002. The Effect of UV Light on the Inactivation of *Giardia lamblia* and *Giardia muris* Cysts as Determined by Animal Infectivity Assay. *Wat. Res.*, 36(8):2098-2108.

Moore, A. M. and C.H. Thompson, 1955. Ultraviolet Irradiation of Pyrimidine Derivatives. *Science*, 122:594.

Morowitz, H.J. 1950. Absorption Effects in Volume Irradiation of Microorganisms. *Science*, 111:229.

Palmateer, G., and E. Whitby. 1987. Ultraviolet Disinfection: Its Effects on *Escherichia coli* and Bacteriophages as Indicators of Disinfection Efficiency of Wastewater. Presented at the Technology Transfer Conference, Royal York Hotel, Toronto, Ont., Canada.

Qualls, R.G., and J.D. Johnson. 1983. Bioassay and Dose Measurement in UV Disinfection. *Appl. Environ. Microbiol.*, 45:872.

Rahn R.O. 1993. Use of Potassium Iodide as a Chemical Actinometer. *Photochem. Photobiol.*, 58:874-880.

Rahn, R.O. 1997. Potassium Iodide as a Chemical Actinometer for 254 nm Radiation: Use of Iodate as an Electron Scavenger. *Photochem. Photobiol.*, 66(4):450-455.

Rahn, R.O., and H.G. Stelin. 1979. Quantum Yield of Photohydration for Dimethyluracil - Concentration and Wavelength Dependence. *Photochem. Photobiol.*, 30:317.

Rice, E.W., and J.C. Hoff. 1981. Inactivation of *Giardia lamblia* Cysts by Ultraviolet Irradiation. *Appl. Environ. Microbiol.*, 42(3):546-547.

Rochelle, P.A., D.M. Ferguson, T.J. Handojo, R. De Leon, M.H. Stewart, and R.L. Wolfe. 1997. An Assay Combining Cell Culture With Reverse Transcriptase PCR to Detect and Determine Infectivity of Waterborne *Cryptosporidium parvum*. *Appl. Environ. Microbiol.*, 63:2029-2037.

Rubin, L.B., T.M. Menshonkova, N.A. Simukova, and E.I. Budowsky. 1981. The Effects of High Intensity UV Radiation on Nucleic Acids and Their Components - I. Thymine. *Photochem. Photobiol.*, 34:339.

Rushizky, G.W., and A.B. Pardee. 1962. Ultraviolet Light Induced Spectral Changes in Nucleotides and Nucleic Acids. *Photochem. Photobiol.*, 1:15.

Scheible, O.K., M. Casey, and A. Forndran. 1986. Ultraviolet Disinfection of Wastewaters From Secondary Effluent and Combined Sewer Overflows. EPA-600/2-86/005. PB86-145182, Cincinnati, Ohio: USEPA.

Shin, G.A., K.G. Linden, and M.D. Sobsey. 1999. Inactivation of *Cryptosporidium* oocysts with Low Pressure UV Radiation. In *Proc., Water Quality Technology Conference, Nov 1-3, 1999, Tampa Bay, Fla.* Denver, Colo.: AWWA.

Shin, G.A., K.G. Linden, and M.D. Sobsey. 2001. Inactivation of Adenovirus by Low Pressure UV Irradiation. Sixth Periodic Report for Fate and Persistence of Pathogens Subjected to Disinfection, Project 98-HHE-02, Water Environment Research Foundation, Alexandria, VA.

Sinsheimer, R.L. 1954. The Photochemistry of Uridylic Acid. *Radiat. Res.*, 1:505-513.

Sinsheimer, R.L., and R. Hastings. 1949. A Reversible Photochemical Alteration of Uracil and Uridine. *Science*, 110:525-526.

Snicer, G.A., J.P. Malley, A.B. Margolin, and S.P. Hogan. 2000. *UV Inactivation of Viruses in Natural Waters*. Denver Colo.: AwwaRF and AWWA.

Sommer, R., and A. Cabaj. 1993. Evaluation of the Efficiency of a UV Plant for Drinking Water Disinfection. *Wat. Sci. Technol.*, 27(3-4):357.

Stolarik G.F., D. Christie, R. Prendergast, T.E.T. Gillogly, and J.A. Oppenheimer. 2001. Long-term Performance and Reliability of a Demonstration-Scale UV Reactor. In P*roc. of First International Congress on Ultraviolet Technologies. June 14-16, 2001, Washington, D.C.* Ontario, Calif.: International Ultraviolet Association.

Swenson, P.A., and R.B. Setlow. 1963. Kinetics of Dimer Formation and Photohydration in Ultraviolet-Irradiated Polyuridylic Acid. *Photochem. Photobiol.*, 2:419.

Thompson, S.S., J.L. Jackson, M. Suva-Castillo, W.A. Yanko, Z.L. Jack, J. Kuo, C.L. Chen, and F.P. Williams. 2000. Detection of Infectious Adenovirus in Tertiary Treated and UV Disinfected Wastewater During a UV Disinfection Pilot Study. In *Proc. WEFTEC 2000, Oct 14-18, 2000, Anaheim, CA*. Alexandria, VA: WEF.

Tosa, K., and T. Hirata. 1998. Photoreactivation of *Salmonella* Following UV Disinfection. In *Proc. IAWQ 19th Biennial International Conference Health-Related Water Microbiology. Vancouver, BC, Canada.*: IWA.

Tree, J.A., M.R. Adams, and D.N. Lees. 1997. Virus Inactivation During Disinfection of Wastewater by Chlorination and UV Irradiation and the Efficacy of F+ Bacteriophage as a 'Viral Indicator'. *Wat. Sci. Tech.*, 35(11-12):227-232.

USEPA (U.S. Environmental Protection Agency). 1986. Design Manual: Municipal Wastewater Disinfection. EPA/625/1-86/021, Cincinnati, OH.: USEPA.

USEPA (U.S. Environmental Protection Agency). 1996a. Ultraviolet Light Disinfection Technology in Drinking Water Application - An Overview. EPA/811/R-96/002, Office of Ground Water and Drinking Water. Washington D.C.: USEPA.

USEPA (U.S. Environmental Protection Agency). 1996. ICR Microbiology Laboratory Manual. EPA/600/R-95/178. Office of Research and Development, Washington, D.C.: USEPA.

USEPA (U.S. Environmental Protection Agency). 1999. Method 1623: *Cryptosporidium* and *Giardia* in Water by Filtration/IMS/FA. EPA-821-R-99-006. Office of Water, Washington, D.C.: USEPA.

von Sonntag, C., and H-P Schuchmann. 1992. UV Disinfection of Drinking Water and By-Product Formation-Some Basic Considerations. *J. Water SRT- Aqua*, 41:67-74.

Wang, S.Y. 1962. Irradiation of Uridine with Ultraviolet Light. *Photochem. Photobiol.*, 1:37- 40.

Webb, R.B. 1977. Lethal and Mutagenic Effects of Near-Ultraviolet Radiation. In *Photochemical and Photobiological Reviews, Vol. 2.* Edited by K.C. Smith. New York: Plenum Press.

Whitby, G.E., G. Palmateer, W.G. Cook, J. Maarschalkerweerd, D. Huber, and K. Flood. 1984. Ultraviolet Disinfection of Secondary Effluent. *J. Wat. Pollut. Control Fed.*, 56:844-850.

Wiedenmann, A., B. Fischer, U. Straub, C.-H. Wang, B. Flehmig, and D. Schoenen. 1993. Disinfection of Hepatitis A Virus and MS-2 Coliphage in Water by Ultraviolet Irradiation: Comparison of UV-Susceptibility. *Wat. Sci. Tech.*, 27(3-4):335-338.

Wilson, B.R., P.F. Roessler, E. Van Dellen, M. Abbaszadegan and C.P. Gerba. 1992. Coliphage MS-2 as a UV Water Disinfection Efficacy Test Surrogate for Bacterial and Viral Pathogens. In *Proc. Water Quality Technology Conference, Nov 15-19, 1992. Toronto, ON, Canada.* Denver, Colo.: AWWA.

Zhang, J.Y., I.W. Boyd, and H. Esrom. 1997. UV Intensity Measurement for a Novel 222 nm Excimer Lamp Using Chemical Actinometer. *Appl. Surf. Sci.*, 109/110:482.

ABBREVIATIONS AND SYMBOLS

Abbreviation	Definition
AwwaRF	Awwa Research Foundation
Cfu	colony-forming unit(s)
cm	centimeter(s)
Crypto.	*Cryptosporidium parvum*
D/DBPR	disinfectants and disinfection byproducts rule
DBP(s)	disinfection byproduct(s)
dia.	diameter
DNA	deoxyribonucleic acid
ES	electrospray
ESI	electrospray interface
ESWTR	Enhanced Surface Water Treatment Rule
ft	foot (feet)
g	gram
g	gravitational force
gal	gallon
Gly-Phe	glycine-phenylalanine
GPD	gallon(s) per day
gpm	gallon(s) per minute
GWDR	Groundwater Disinfection Rule
h	hour
HAA(s)	haloacetic acid(s)
HPC	heterotrophic plate count
hr(s)	hour(s)
IFA	immuno fluorescence assay
in.	inch(es)
J	joule(s)
kg	kilogram(s)
km	kilometer(s)
kW•hr	kilowatt-hour(s)

L	liter(s)
LC	liquid chromatography
LT2ESWTR	long-term 2 ESWTR
m	meter(s)
M/DBPR	microbial and disinfection byproducts rules
m/z	mass-to-unit charge ratio
mg/L	milligram(s) per liter
mgd, MGD	million gallon(s) per day
min	minutes
mJ	millijoule(s)
mL	milliliter(s)
mm	millimiter(s)
mM	millimolar
MPN	most probable number
MS	mass spectrometer
mW	milliwatt(s)
μm	micrometer(s)
μW	microwatt(s)
n	number of samples
N	concentration of microorganism
nm	nanometer(s)
No	initial concentration of microorganism
NOM	natural organic matter
ntu, NTU	nephelometric turbidity unit(s)
O.D.	outside diameter
PBS	phosphate-buffer solution
pfu	plaque-forming unit(s)
pH	negative logarithm of the effective hydrogen-ion concentration
ppm	parts per million
psi	pounds per square inch
Q	flow
QA/QC	quality assurance / quality control
R	correlation coefficient
RNA	riboneucleic acid
RO	reverse osmosis
s	second(s)
S	surface area
SDWA	Safe Drinking Water Act
SIR	selected ion recording

sq. cm	square centimeter(s)
sq in.	square inch(es)
SWTR	Surface Water Treatment Rule
T	temperature
THM(s)	trihalomethane(s)
TOC	total organic carbon
TOF	time of flight
US	United States
USEPA	United States Environmental Protection Agency
UV	ultraviolet
UV_{254}	Absorbance of 254-nanometer ultraviolet light
UVA	ultraviolet light from 315 – 400 nm
UVB	ultraviolet light from 280 – 315 nm
UVC	ultraviolet light from 200 – 280 nm
v	velocity
V	volt(s)
W	watt(s)

Symbol	**Definition**
@	at
°C	degree(s) Celsius
°F	degree(s) Fahrenheit
%	percent
±	plus and minus standard deviation

IWA Publishing's authorised EU representative for General Product Safety Regulations is Diane D'Arras, 15 rue Duret, 75116 Paris, France, e-mail: safety@iwap.co.uk.

Printed and bound by CPI Group (UK) Ltd, Croydon, CR0 4YY
23/09/2026
02221720-0005